21世纪高职高专精品教材·英语系列

能源电力专业基础英语

BASIC ENGLISH FOR ENERGY AND ELECTRICAL POWER

主　编　黄月花　苏　慧
副主编　吴玥偎　陆康东
编　者　李尚凤　项　霄　李含霜

中国人民大学出版社
·北京·

图书在版编目（CIP）数据

能源电力专业基础英语 / 黄月花，苏慧主编. — 北京：中国人民大学出版社，2019. 10
21世纪高职高专精品教材. 英语系列
ISBN 978-7-300-27532-1

Ⅰ. ①能… Ⅱ. ①黄… ②苏… Ⅲ. ①能源工业–英语–高等职业教育–教材 ②电力工业–英语–高等职业教育–教材 Ⅳ. ①TK01 ②TM

中国版本图书馆 CIP 数据核字（2019）第 222152 号

21 世纪高职高专精品教材 · 英语系列
能源电力专业基础英语
主　编　黄月花　苏　慧
副主编　吴玥偎　陆康东
编　者　李尚凤　项　霄　李含霜
Nengyuan Dianli Zhuanye Jichu Yingyu

出版发行	中国人民大学出版社		
社　址	北京中关村大街 31 号	邮政编码	100080
电　话	010–62511242（总编室）		010–62511770（质管部）
	010–82501766（邮购部）		010–62514148（门市部）
	010–62515195（发行公司）		010–62515275（盗版举报）
网　址	http:// www. crup. com. cn		
经　销	新华书店		
印　刷	北京宏伟双华印刷有限公司		
规　格	185 mm × 260 mm 16 开本	版　次	2019 年 10 月第 1 版
印　张	7.75	印　次	2022 年12月第 4 次印刷
字　数	185 000	定　价	37.00 元

前 言
PREFACE

《能源电力专业基础英语》是课题《在高职基础英语教学中融入专业英语的研究与实践》的研究成果，由课题主持人及课题组成员编写而成，课题组成员有数十年的教学经验。本教材是课题研究的主要成果之一，融合了基础英语和专业英语的内容，在高职基础英语中融入能源电力专业基础知识。

一、编写原则

教育部关于《以就业为导向深化高等职业教育改革的若干意见》之高职教育的特点：应用性、实践性和实用性；2014 年国务院《关于加快发展现代职业教育的决定》(以下简称《决定》) 中指出“加快现代职业教育体系建设，深化产教融合、校企合作，培养数以亿计的高素质劳动者和技术技能人才。”根据以上的要求和《决定》中的总体要求，本教材遵循了以下原则：

1. 语言技能与专业知识相结合

根据学生的英语学习特点，本教材编写时在英语语言技能知识训练的基础上，扩展了专业知识，注重在语言训练中融入基本的专业知识和内容，兼顾了英语语言技能训练与专业知识的学习。

2. 教材凸显教育实际运用、图文并茂

本教材联系生活实际，在进行基础语言技能训练的同时加入了能源电力专业基础知识，文中配有与专业相关的图片、流程图，让学生在视觉冲击中理解抽象内容，同时教材也关注学生的英语应用能力。具体如下：

(1) 每单元都有模块化的主题，围绕主题进行听、说、读、写、译的技能训练，以丰富对专业知识的认知 (每单元的第一部分是根据关键词进行专业知识导入，一般为相关问题和图片)。

(2) 教材配有图片、设备图、流程图，可以引起视觉冲击从而吸引学生的兴趣，帮助学生理解抽象的阅读内容，从而在语言环境中加深对专业知识的学习。(如第一章配有电力工程组成图；第二章配有电力系统组成要素图；第三章配有电厂工艺流程图和设备图；第四章配有火力发电厂生产过程图；第五章配有水力发电厂工作流程图；第六章配有太阳能发电厂的流程图；第七章配有输配电流程图等)

(3) 每单元课后设置了词汇练习，如词性变换题、选词填空、句子翻译等，学生可以运用语言工具逐步深入学习并掌握新词汇，既兼顾了基础词汇学习，也拓展了专业词汇学习，如 generate—发生〈基础词汇〉, generator—发电机〈专业词汇〉。

(4) 本教材融合了“高等学校英语应用能力考试”的内容 (如课后练习中的阅读练习、词汇语法练习、翻译练习、应用文写作等)，符合逐步提高学生语言认知和应用能力的要求。

二、教材内容结构

本书每单元涉及一个主题专业内容，并围绕这个专业内容进行语言的听、说、读、写、译的内容设计和编排，融合了能源电力专业基础知识。本书每单元分成 Section A Professional Background——Lead-in、Listening & Speaking，Section B Text Learning——Lead-in、Text A、Text B，Section C Advanced Training——Grammar、Writing，Section D An English Song 8 个部分，共 7 个单元主题。

Unit 1 Electrical Power Engineering 电力工程 教学重点：Lead-in 通过图片了解关于电的词汇和表达；Dialogues 理解主题对话，填写缺少的单词；Text A 理解课文内容，掌握电力工程的基本知识，Text B 为知识拓展，掌握相关的词汇；Grammar 掌握语法：形容词；Writing 掌握英语写作：会写通知；听歌学英语。

Unit 2 Electrical Power System 电力系统 教学重点：Lead-in 通过图片了解关于电网、电力系统的词汇和表达；Dialogues 理解主题对话，填写缺少的单词；Text A 理解课文内容，掌握电力系统的基本知识，Text B 为知识拓展，掌握相关的词汇；Grammar 掌握语法：谓语动词；Writing 掌握英语写作：会写邀请信；听歌学英语。

Unit 3 Power Plant Equipment 电厂设备 教学重点：Lead-in 通过图片了解关于电厂的常见设备词汇和表达；Dialogues 理解主题对话，填写缺少的单词；Text A 理解课文内容，掌握电厂的基本组成部分及设备的使用，Text B 为知识拓展，掌握相关的词汇；Grammar 掌握语法：英语基本句型；Writing 掌握英语写作：会写感谢信；听歌学英语。

Unit 4 Thermal Power Plant 火电厂 教学重点：Lead-in 通过图片了解关于火电厂的设备词汇和表达；Dialogues 理解主题对话，填写缺少的单词；Text A 理解课文内容，掌握火电厂的基本组成部分及设备的运行，Text B 为知识拓展，掌握相关的词汇；Grammar 掌握语法：动词的时态；Writing 掌握英语写作：会写邀请函；听歌学英语。

Unit 5 Hydropower Plant 水电厂 教学重点：Lead-in 通过图片了解关于水电厂的词汇和表达；Dialogues 理解主题对话，填写缺少的单词；Text A 理解课文内容，掌握水电厂的基本组成部分及设备的运行，Text B 为知识拓展，掌握相关的词汇；Grammar 掌握语法：动名词；Writing 掌握英语写作：会写备忘录；听歌学英语。

Unit 6 Concentrated Solar Power 聚光太阳能发电 教学重点：Lead-in 通过图片了解关于太阳能发电的词汇和表达；Dialogues 理解主题对话，填写缺少的单词；Text A 理解课文内容，掌握太阳能发电的使用及相关知识，Text B 为知识拓展，掌握相关的词汇；Grammar 掌握语法：动词分词；Writing 掌握英语写作：会写申请表；听歌学英语。

Unit 7 Transmission and Distribution Systems 输配电系统 教学重点：Lead-in 通过图片了解关于输配电的词汇和表达；Dialogues 理解主题对话，填写缺少的单词；Text A 理解课文内容，掌握输配电的相关知识，Text B 为知识拓展，掌握相关的词汇；Grammar 掌握语法：动词不定式；Writing 掌握英语写作：会写简历；听歌学英语。

本教材可以供高职高专能源电力相关专业学生第一学期使用，也可以第二学期使用，为专业英语学习打下基础。本教材的编写为线上平台搭建提供了内容。当然，限于作者水平，如有疏漏之处，请广大师生批评指正。

本书配有录音文件和参考答案，请扫描封底二维码下载，或登录中国人民大学出版社官方网站 http://www.crup.com.cn 搜索本书后下载，或联系出版社免费索取：010-62515580、010-62513365；chengzsh@crup.com.cn。

编者

2019 年 8 月

目　录

CONTENTS

Unit 7 Transmission and Distribution Systems 输配电系统

Unit 1

Electrical Power Engineering

电力工程

Section A Professional Background

Part One Lead-in

☞ **Task** Study the pictures and discuss the questions below.

Questions:

1. What can you see in these pictures?
2. What are they used for?

Cues:
bulb, coal, LED tube, power station, substation, lines, power pole, tower, transformer, mountain

Part Two Listening & Speaking

Listen to the following dialogues and fill in the blanks.

☞ **Dialogue 1**

W: Tom, would you like going to see the film with me? I have two __________ for *The 7 Adventures of Sinbad* (辛巴达七次历险) at 1 p.m..

M: I am sorry, I am __________ this morning.

W: What are you busy with, Tom? You are __________ the film for a long time.

M: I have to __________ my homework. Professor Li asked us to hand it in tomorrow morning.

W: You can do it in the evening.

M: According to the notice, the electricity will __________ tonight.

W: It is inconvenient without electricity.

M: Yes, I can't __________ how to do without electricity.

☞ **Dialogue 2**

W: Hey Rob, what is your major?

M: I major in Centralized Control __________ of Thermal Power Plant (火电厂集控运行).

W: Oh, it's a ___________ major. You can work in the power plant after graduation.

M: Yes, I want to get into the Guangxi Laibin Power Plant B (广西来宾 B 电厂).

W: Do you know the power plants in Guangxi?

M: We have the steam power plant, hydropower plant, nuclear power plant.

W: Except these three types of power plants, are there other types of power plants?

M: Yes, there are solar, ___________, tidal and geothermal power plants.

W: They are ___________ new energy ___________.

Section B　Text Learning

Part One　Lead-in

1. Do you know who is the creator of electric bulb?
2. What areas can electricity be used for?
3. How can we use electricity at home?

Part Two

☞ Text A

Introduction to Electrical Power Engineering

Because of its easy conversion, transmission and control, electric power has gradually replaced steam power since the power engineering generation in the 1880s. After the 20th century, the production of electricity mainly depended on thermal power plants (steam power plants), hydropower plants and nuclear power plants. There are still other new energy resources for electricity generation, such as tidal power, solar power, geothermal power and wind power. The transmission and distribution of electric energy is mainly realized through the high and low voltage AC power network. As the direction of the development of transmission engineering technology, the emphasis is to study the AC transmission and DC transmission technology of ultra-high voltage (over one million volt) and to form a larger power network.

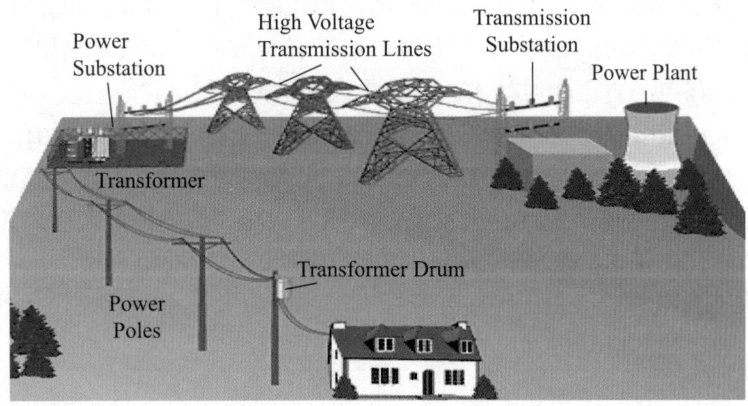

Fig. 1.1

Electricity generation, transmission, and distribution

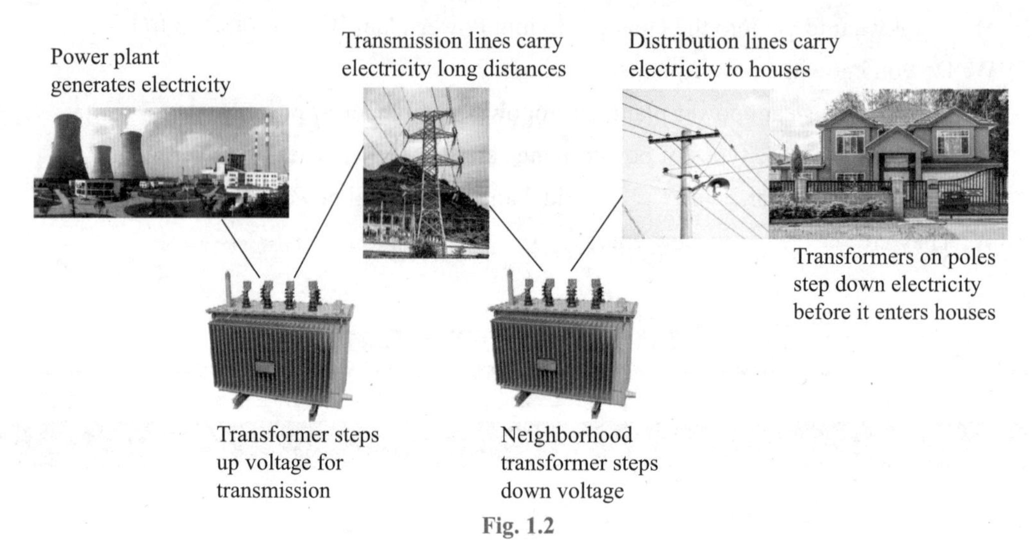

Fig. 1.2

Power engineering, also called power systems engineering, is a subfield of electrical engineering that deals with the generation, transmission, distribution and utilization of electric power, and the electrical apparatus connected to such systems. As a product, electricity is different from other types of products because it cannot be stored easily. Therefore, the power system composed of power plant, transmission line, substation and distribution network must be in accordance with the consumption of the user's electricity.

Power engineering deals with the generation, transmission, distribution and utilization of electricity as well as the design of a range of related devices. These include transformers, electrical generators, electric motors and power electronics.

(224 words)

参考译文

电力工程

电能因易于转换、传输、控制，从19世纪80年代电力工程发电以后，已逐步取代蒸汽动力。20世纪以后，电能的生产主要靠热电厂（火力发电厂）、水电站和核电站。还有新能源发电比如潮汐能、太阳能、地热能和风能。电能的输送和分配主要通过高、低压交流电力网络来实现。作为输电工程技术发展的方向，其重点是研究特高压（100万伏以上）交流输电与直流输电技术，形成更大的电力网络。

电力工程，又称电力系统工程，是电气工程的一个分支，涉及电力的产生、传输、分配和利用，以及连接到这些系统的电气设备。电能作为一种产品，和其他类型的产品不同是因为它不易于储存。因此，由发电厂、输电线路、变电所和配电网组成的电力系统每时每刻所生产、输送的电能，都必须和用户电能的消费量相一致。

电力工程涉及电力的产生、传输、分配和利用，以及一系列相关设备的设计。其中包括变压器、发电机、电动机和电力电子设备。

Words & Expressions	
conversion [kən'vɜːʃn]	*n.* 转变；换算
transmission [træns'mɪʃn]	*n.* 传输；传播；变速器
electricity [ɪˌlek'trɪsəti]	*n.* 电
voltage ['vəʊltɪdʒ]	*n.* 电压
apparatus [æpə'reɪtəs]	*n.* 设备；装置；仪器
transformer [træns'fɔːmə(r)]	*n.* 变压器
generator ['dʒenəreɪtə(r)]	*n.* 发电机；发生器
electronics [ɪˌlek'trɒnɪks]	*n.* 电子学；电子器件
thermal power plant (steam power plant)	热电厂（火力发电厂）
hydropower plant	水电站
nuclear power plant	核电厂
tidal power	潮汐能
solar power	太阳能
geothermal power	地热能
wind power	风能
electric motor	电动机

Notes

1. deal with: 涉及；处理
   e.g. This matter dealt with the profits of the company. 这件事涉及公司的利益。
2. in accordance with: 与……一致；依照　(be in accordance with)
   e.g. He said the company would deal with the matter in accordance with articles of the company. 他表示，公司将根据公司条款处理此事。

Exercises

☞ **Task I**　Answer the following questions according to the text.

1. Why has the electric power replaced steam power?
   ____
2. What does the production of electricity mainly depend on after the 20th century?
   ____
3. How is the transmission and distribution of electric energy mainly realized?
   ____
4. What is power engineering?
   ____
5. What does the power system compose of?
   ____

☞ **Task 2** Change the word formation.

Before	After	Meaning
distribution *n.*		*vt.* 分配；散布；分开；把……分类
transmission *n.*		*vt.* 传达；传染；传送 *vi.*（以无线电或有线电的方式）发送信号
electricity *n.*		*adj.* 电子的；电子学的
voltage *n.*		*n.* 伏特
utilization *n.*		*vt.* 〈美〉利用或使用=〈英〉utilise

☞ **Task 3** Complete the following sentences with the proper form of the words from Task 2.

1. I want to __________ the new energy.
2. I like to read __________ books.
3. The unit of __________ is the volt.
4. Power stations are __________ among many areas in China.
5. Please __________ my best regards to your family.

☞ **Task 4** Translate the following sentences into Chinese.

1. After the 20th century, the production of electricity mainly depended on thermal power plants, hydropower stations and nuclear power plants.

____

____

2. There are still other new energy resources for electricity generation, such as tidal power, solar power, geothermal power and wind power.

____

____

3. Power engineering, also called power systems engineering, is a subfield of electrical engineering.

____

____

4. Electricity is different from other types of products because it cannot be stored easily.

____

____

5. Therefore, the power system is composed of power plant, transmission line, substation and distribution network.

____

____

Part Three

☞ **Text B**

The Latest Trends in Energy Development

The steady development of China's economy has brought the strong growth in energy demand, not only to promote the great development of China's energy industry and the transformation of the structure, but also make China's new energy industry in the top in the world. At the same time, China's economic and social development make people have an urgent need to improve the ecological environment. "Pollution prevention" at the end of 2017 is one of the three big battles in the next three years. This demand also greatly promotes the revolution of China's energy consumption, and makes the government speed up the transformation of the institutions for energy consumption. With the further expansion of opening to the outside world as well as the implementation of the "Belt and Road" initiative, China is set to a more in-depth participation in the process of global economic governance. We have reason to believe that China will play a more active and important role in the world economic development and the international energy market, and vitalize the development of the global economic governance.

(177 words)

参考译文

能源发展的最新趋势

中国经济的稳定发展带来了能源需求的强劲增长，不仅推动了中国能源产业的巨大发展和结构的转型，而且使中国新能源产业走在了世界前列。同时，中国经济社会的发展使人们对生态环境的改善有了更迫切的诉求，2017 年底，“污染防治”被列为未来三年三大攻坚战之一。这种需求也极大地推动了中国的能源消费革命，使中国政府更加着力加快能源消费机构的转型发展。随着对外开放的进一步扩大以及“一带一路”倡议的实施，中国势必将更加广泛深入地参与全球经济治理的进程。我们有理由相信，中国在世界经济发展和国际能源市场中将发挥更加积极、更加重要的作用，为全球经济治理的发展注入更多鲜活动力。

Words & Expressions

promote [prə'məʊt]	*vt.* 促进，推进；提升
ecological [ˌiːkə'lɒdʒɪkl]	*adj.* 生态（学）的
battle ['bætl]	*n.* 战争，战役
revolution [ˌrevə'luːʃn]	*n.* 革命；彻底改变
consumption [kən'sʌmpʃn]	*n.* 消费；消耗
expansion [ɪk'spænʃn]	*n.* 扩大；扩张
participation [pɑːˌtɪsɪ'peɪʃn]	*n.* 参加，参与；分享
governance ['gʌvənəns]	*n.* 统治；管理；支配；统治方式

vitalize ['vaɪtəlaɪz]	*vt.* 赋予生命；给予……生命；使有生气
pollution prevention	污染防治
Belt and Road	“一带一路”

Notes

1. have an urgent need: 有紧急需要，迫切需求
   e.g. They have an urgent need to import advanced technology and equipment.
   他们在先进的技术和设备引进方面有迫切的需求。
2. with the expansion of: 对……进一步扩大
   e.g. These could not keep up with the expansion of the factory.
   这些都跟不上工厂的发展。

Exercises

☞ **Task I** The following pictures are about saving energy. Fill in the blanks with the right words or phrases according to the marks in the picture.

① ______________________________
② ______________________________
③ ______________________________
④ ______________________________
⑤ ______________________________
⑥ ______________________________

☞ **Task 2** Match the words below with their definitions.

1. promote	赋予生命；给予……生命；使有生气
2. battle	促进，推进；提升
3. expansion	统治；管理；支配；统治方式
4. governance	战争，战役
5. vitalize	扩大；扩张

☞ **Task 3** Translate the following sentences into Chinese.

1. The steady development of China's economy has brought the strong growth in energy demand.

____

____

2. At the same time, China's economic and social development make people have an urgent need to improve the ecological environment.

____

____

3. "Pollution prevention" at the end of 2017 is one of the three big battles in the next three years.

____

____

4. With the further expansion of opening to the outside world as well as the implementation of the "Belt and Road" initiative, China is set to a more in-depth participation in the process of global economic governance.

____

____

5. China will play a more active and important role in the world economic development and the international energy market.

____

____

Section C Advanced Training

Part One Grammar: Adjective 形容词

一般以 -able, -ful, -al, -ic, -less, -ish, -ous, -y, -ed, -ive 为后缀的词是形容词。形容词主要用来描写或修饰名词或代词，形容词用于表示人或事物的性质、状态、特征或属性，常用作定语，也可作表语、补语或状语。

1. 用作定语（可以放名词前面作前置定语，也可放名词后面作后置定语）。

e.g. This is a useful machine. 这是一台有用的机器。（前置定语）

This is a girl dressed in green. 这是一个穿绿色衣服的女孩。（后置定语）

2. 用作表语（形容词作表语，放在系动词之后）。

e.g. The maintenance costs of transmission lines are very expensive. 输电线路的维护费用很高。

备注：有些动词常接形容词作表语。

（1）become, come, fall, get, go, grow, make, turn（表示“变成某种状态”）

e.g. He makes his job become interesting. 他把自己的工作变得很有趣。

（2）continue, hold, keep, lie, remain, stay（表示“保持某种状态”）

e.g. We should exercise everyday to keep healthy. 我们应该每天锻炼来保持身体健康。

（3）appear, feel, look, smell, sound, taste, know（表示“感觉”）

e.g. He feel nervous at the first day of work. 他第一天上班感到紧张。

3. 用作宾语补足语（形容词作宾补，放在宾语之后，与之构成复合宾语）。

e.g. He always keeps the window open. 他总开着窗户。

4. 某些形容词前加定冠词表示一类人或物，为复数概念，在句中起名词作用，可作主语或宾语。e.g. the old

5. 某些以 -ly 结尾的词不是副词而是形容词，friendly, lonely, lovely, likely, daily, lively 等。

e.g. she is a lively girl. 她是一个可爱的女孩。

Exercises

Choose the appropriate answer from the three choices marked A, B and C.

1. Last night, Taiwan oil refineries exploded. This is an ___________ news.

   A. amazing　　B. amazed　　C. amaze

2. This kind of pole looks ___________.

   A. durables　　B. durable　　C. durably

3. He became ___________ when he heard the news.

   A. excites　　B. exciting　　C. excited

4. I like the contestant ___________ in red.

   A. dressing　　B. dressed　　C. dress

5. Which is ___________ hydropower station, Yantan, Tianshengqiao（天生桥）or Longtan?
   A. big　　B. bigger　　C. the biggest
6. The ___________ transmission capacity is 10,000 KWH.
   A. day　　B. daily　　C. days

Part Two　Writing

Notice 通知

通知是上级对下级、组织对成员或学校、系部对学生用于部署工作、传达事物或召开会议等的知照性公文。发通知方与被通知方都以第三人称出现。

（1）写通知一般应注意以下几点

① 通知通常在正文上方的正中间写上 Notice 或 NOTICE（通知）。

② 内容必须包括具体时间、地点、事宜和通知对象等，表达必须准确，措辞也要较正式。

③ 必须包括写通知的个人或者单位的称谓，而且一般都放在通知的右下角。必须包括写通知的日期，日期放在通知的个人或单位的称谓的下一行右下角。

（2）常用句型

A meeting on basketball match will be held at 5 : 30 p.m. on June 22 (Friday) in the meeting room.

[译文] 将于 6 月 22 日（星期五）下午 5 : 30 在会议室召开有关篮球比赛的会议。

All the monitors are requested to be present on time.

[译文] 所有班长必须准时出席。

（3）通知例文

说明： 请以能源动力与发电工程系学生会的名义给本系各班长发一则通知。

内容： 能源动力与发电工程系本月将举行篮球比赛，所有班长必须在 2018 年 6 月 25 日（周五）下午 4 : 30 到 1-201 会议室参加会议，讨论篮球比赛事宜。

Sample

Notice

Our department will hold a basketball match during this month. All the monitors are required to attend the meeting at 4:30 p.m. on June 25 (Friday) in Conference Room 1-201 to discuss the match.

Students' Union of Energy Dynamics and Power Generation Engineering Department

Jun 20, 2018

☞ **Task**

说明： 以电力工程系的名义于 2018 年 6 月 15 日发一则通知，内容如下：2017 级发电厂及电力系统专业 1701 班和 1702 班将于 2018 年 6 月 20 日早上 8 : 30 分在图书馆前坐车去麻石电厂参观，请两个班的同学按时上车。如有紧急情况无法前往，请与各班班长联系。

各班班长做好各班人数的统计。

Words for reference：参观 visit　　按时 on time　　紧急情况 emergency

Section D　An English Song

☞ **Task**　Listen to the song and fill in the blanks with the missing words you have just heard, and then sing along with it.

The Power of the Dream

Celine Dion

Deep within each heart 在每个人心灵深处
There lies a magic spark 都蕴藏着神奇的火花
That lights the fire of our imagination 它点燃我们的想象之火
And since the dawn of man 自从人类降生
The strength of just I can 它就是我独有的力量
Has brought together people of all nations 将全世界的人们聚在一起
There's nothing ordinary 没有人是平凡的
In the living of each day 在每一天的生活里
There's a special part 有一个与众不同的角色
Every one of us will play 我们每个人都会扮演
Feel the flame forever burn 感觉那永远燃烧的火焰
① ______________________________

To bring us closer to the power of the dream 使我们与梦想的力量更近
As the world gives us its best 世界给予我们它最好的精华
To stand apart from all the rest 使我们超然于芸芸万物
It is the power of the dream that brings us here 是梦想的力量把我们聚到了一起
Your mind will take you far 意志会带你走更远
The rest is just pure heart 仅存的就是纯净的内心
You'll find your fate is all your own creation 你会发现你的命运由你自己创造
Every boy and girl 每个男孩和女孩
② ______________________________

They bring the gift of hope and inspiration 同时带来了希望与灵感的天赋
Feel the flame forever burn 感觉那永远燃烧的火焰
Teaching lessons we must learn 教授着我们必须学习的课程
To bring us closer to the power of the dream 使我们与梦想的力量更近

The world unites in hope and peace 世界是希望与和平的结晶
③ ______________________________

It is the power of the dream that brings us here 是梦想的力量把我们带到这里
There's so much strength in all of us 我们有那么大的力量
Every woman, child and man 每个女人 孩子和男人
It's the moment that you think you can't 当你认为你不行的时候
You'll discover that you can 你会发现你可以
Feel the flame forever burn 感觉那永远燃烧的火焰
Teaching lessons we must learn 教授着我们必须学习的课程
To bring us closer to the power of the dream 使我们与梦想的力量更近
The world unites in hope and peace 世界是希望与和平的结晶
We pray that it will always be 我们祈祷它永远都是这样
It is the power of the dream that brings us here 是梦想的力量把我们带到这里
Feel the flame forever burn 感觉那永远燃烧的火焰
Teaching lessons we must learn 教授着我们必须学习的课程
To bring us closer to the power of the dream 使我们与梦想的力量更近
The world unites in hope and peace 世界是希望与和平的结晶
We pray that it will always be 我们祈祷它永远都是这样
It is the power of the dream that brings us here 是梦想的力量把我们带到这里
The power of the dream 梦想的力量
④ ______________________________

The courage to embrace your fear 直面恐惧的力量
No matter where you are 不管你在哪里
To reach for your own star 你要成为自己的明星
To realize the power of the dream 去发挥梦想的力量

Background Tip:

席琳·迪翁（Celine Dion），1968 年 3 月 30 日生于加拿大魁北克省，加拿大歌手。1980 年，12 岁的席琳·迪翁步入歌坛，15 岁时推出首支法文单曲，1990 年推出首张英文专辑 *UNISON*。1996 年为美国亚特兰大奥运会演唱了主题曲 *The Power of the Dream*。1997 年为电影《泰坦尼克号》献唱片尾曲 *My Heart Will Go On*，并获得第 70 届奥斯卡最佳电影歌曲奖。

Unit 2

Electrical Power System

电力系统

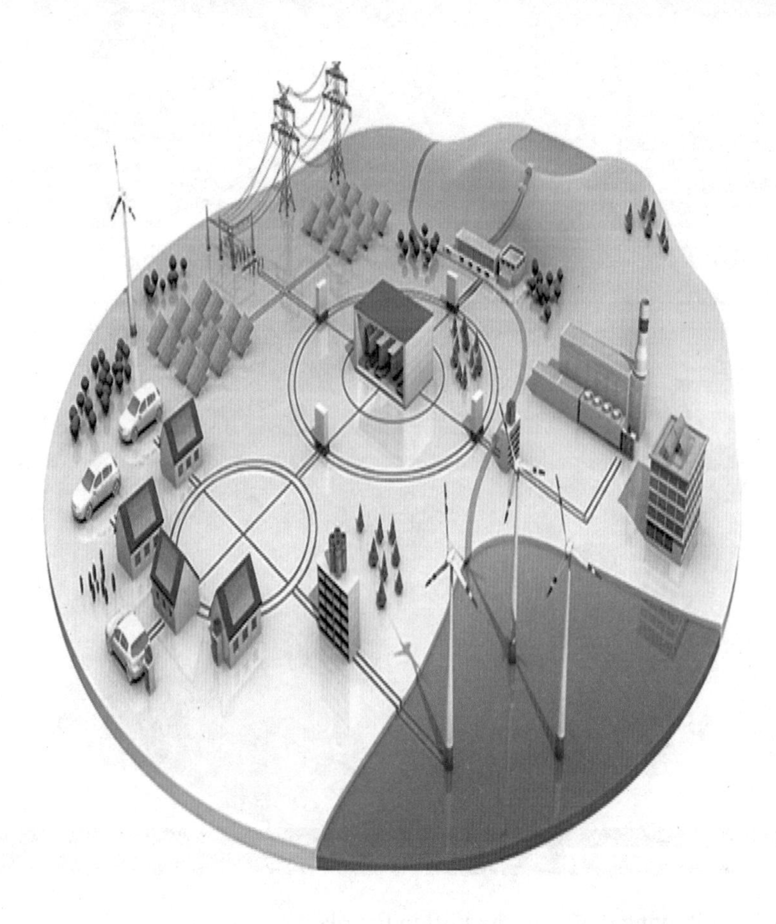

Section A Professional Background

Part One Lead-in

☞ **Task** Study the pictures and discuss the questions below.

Questions:

1. What can you see in these pictures?
2. What is the Electric Grid?

Cues:
generation, transmission, distribution, consumption, substation, solar cell, wind power, residential, thermal power, smart grid

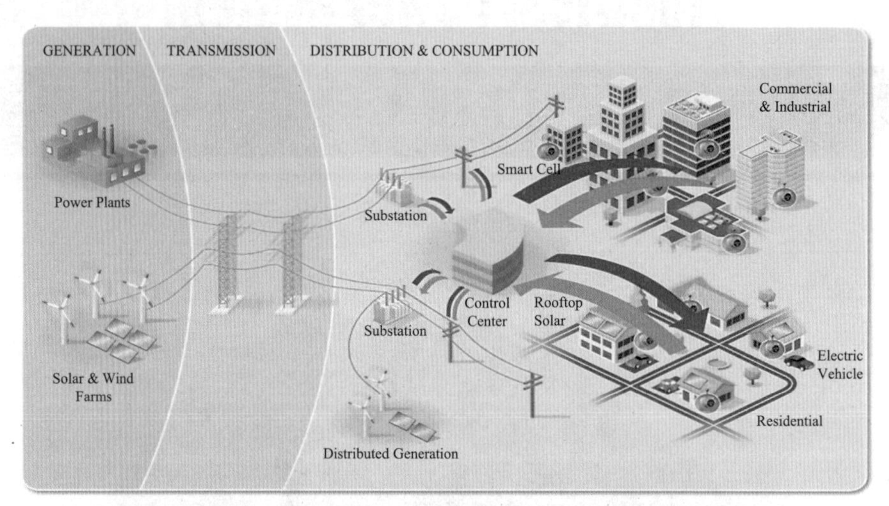

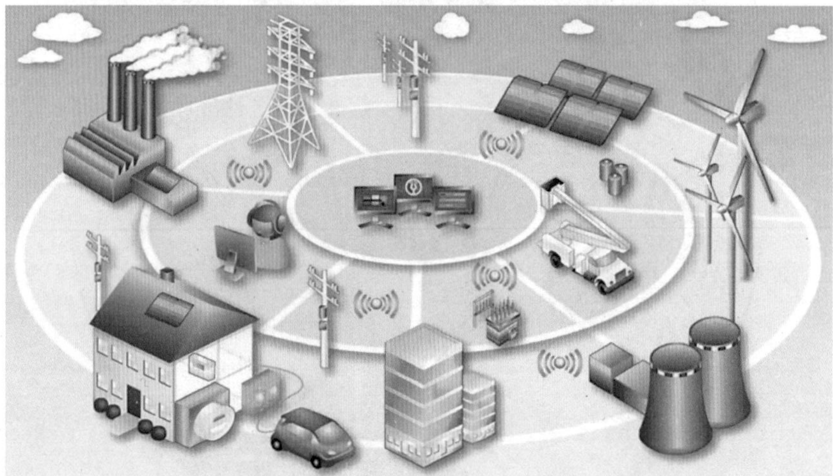

Part Two Listening & Speaking

Listen to the following dialogues and fill in the blanks.

Dialogue I

W: Bob, are you a freshman?

M: Yes, I am in the Department of Electrical Engineering.

W: What major do you ___________?

M: Power Station and Power System.

W: Can you ___________ me how many majors are there in your department?

M: There are about six majors.

W: What is the most ___________ major?

M: Power Station and Power System is a ___________ major in Guangxi.

W: Oh, I ___________.

☞ **Dialogue 2**

M: Hey Jenny, I heard that your ___________ is in the Guangxi Power Grid Company.

W: Yes, my father is one of the ___________.

M: Can you tell me what the electrical power system is?

W: An electrical power system is a ___________ of electrical components deployed to ___________, transfer, distribute and ___________ electric power.

M: Oh, I see.

W: When I was a kid, my father took me to see the substation.

M: I am also eager to see the substation and its operation.

W: I think you will have the chance.

Section B　Text Learning

Part One　Lead-in

1. What are the elements of an electrical power system?
2. Do you know the power stations in Guangxi? What are they?

Part Two

☞ **Text A**

Electrical Power System

A steam turbine is used to provide electric power. An electrical power system (Fig. 2.1) is a network of electrical components deployed to supply, transfer, distribute and use electric power. An example of an electrical power system is the grid . "Electric Grid" basically describes a complete network which includes transmission lines, transformers, distribution substation and all accessories that are used for delivery of electricity from generation plants to home and commercial scale. An electrical grid power system can be broadly divided into the generators that supply the power, the transmission system that carries the power from the generating centers to the load centers (consumers), and the distribution system that feeds the power to nearby homes and industries. The most usual system today for generation and for the general transmission of power is the three-phase system.

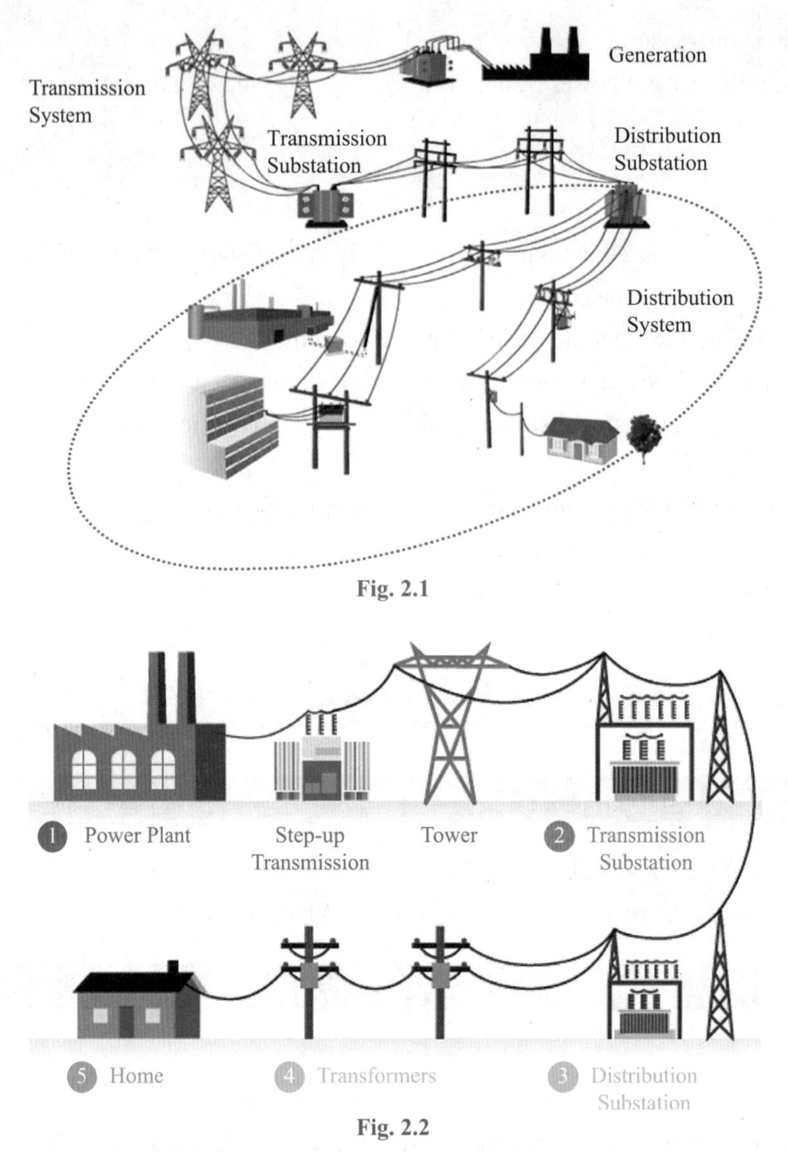

Fig. 2.1

Fig. 2.2

In fact, a large system includes numerous generating stations, transmission system, distribution system and loads (consumers). Electricity is generated from the generating stations like steam power stations (coal, oil or natural gas), hydroelectric power stations, nuclear power stations, wind power stations, solar power stations and so on. Over a long distance before transmission, electricity is stepped up at high voltage by transformers in the step-up substation. Then it is carried by the transmission line. Before electricity is used by consumers, the voltage must be stepped down by the step-down transformer. The electricity is transferred through the distribution system.

(230 words)

参考译文

电力系统

汽轮机用于提供电力。电力系统（图 2.1）是一个由电气元件组成的网络，用于供应、传输、分配和使用电力。以电力系统中的电网为例。“电网”基本上描述了一个完整的网络，包括输电线路、变压器、配电变电站、各种辅助装置，它们一起把电从发电厂输送到家庭和商业用户。电网电力系统可以大致分为供电的发电机，将发电中心的电力输送到负荷中心（消费者）的输电系统，以及将电力输送到附近家庭和工业的配电系统。当今最常见的发电和输电系统是三相系统。

事实上，一个大系统包括许多发电站、输电系统、配电系统和负荷（消费者）。电力是由火力发电站（煤、石油或天然气）、水电站、核电站、风力发电站、太阳能发电站等发电站产生的。在传输前的一段很长的距离内，在升压变电站变压器将电压升高，然后由传输线传送。在用户使用电力之前，降压变压器将电压降低，电力通过配电系统传输。

Words & Expressions

steam turbine	*n.* 汽轮机，蒸汽轮机
transfer [træns'fɜː(r)]	*v.* 转移；调任；转乘 *n.* 迁移；移动；换车；汇兑
store [stɔː(r)]	*n.* 商店；贮藏 *vt.* 储存；贮藏；保存
grid [grɪd]	*n.* 电网；网格
generator ['dʒenəreɪtə]	*n.* 发电机，发生器；生产者
generating centre	发电中心
generating station	发电站
load center	负荷中心
feed [fiːd]	*vt.* 喂养；饲养；向……提供
three-phase [θ'riːf'eɪz]	*adj.* 三相的 *n.* 三相
step-up	*adj.* 把电压升高的
step-up substation	升压变电站
transformer [træs'fɔːmə(r)]	*n.* 变压器
distribution system	配电系统

Notes

1. electrical grid power system: 电网电力系统

   e.g. 中国国家电网 State Grid Corporation of China

   电网：在电力系统中，连接发电和用电的设施和设备的统称。属于输送和分配电能的中间环节，它主要由联结成网的送电线路、变电所、配电所和配电线路组成。

电力系统：由发电、输电、变电、配电和用电等环节组成的电能生产、传输、分配和消费的系统。

2. three-phase system: 三相系统　　三相交流电是电能的一种输送形式
3. step-up substation: 升压变电站　　distributing substation: 配电站
4. step-down transformer: 降压变压器

Exercises

Task 1 Read the text and judge the following statements with true (T) or false (F).

(　　) 1. A steam turbine is used to provide electric power.

(　　) 2. An electrical power system is a network of electrical components deployed to supply, transfer, distribute and use electric power.

(　　) 3. The most usual system today for generation and for the general transmission of power is the single-phase system.

(　　) 4. Electricity is stepped down at low voltage by transformers in the step-up substation.

(　　) 5. The electricity is transferred through the generation system.

Task 2 Change the word formation.

Before	After	Meaning
transfer *v.*		*adj.* 可转移的；可转让的
store *v.*		*adj.* 存信息的；存储的
generator *n.*		*vt.* 产生；发生；引起
transformer *n.*		*vt.* 改变；转换 *vi.* 改变；变形
distribution *n.*		*v.* 分配；散发；分布

Task 3 Complete the following sentences with the proper form of the words from Task 2.

1. We can't ___________ enough power for the entire city.
2. Electricity can ___________ by the transmission line.
3. The voltage of electricity must be ___________ during the transmission and distribution system.
4. The teacher ___________ books among the students.
5. Electricity can't be ___________ easily because it is different from other products.

Task 4 Translate the following sentences into Chinese.

1. A steam turbine is used to provide electric power.

____

____

2. An electrical power system is a network of electrical components.

____

____

3. Electricity is generated from the generating stations.

____

____

4. In fact, a large system includes numerous generating stations, transmission system, distribution system and loads (consumers).

____

____

5. Over a long distance before transmission, electricity is stepped up at high voltage by transformers in the step-up substation.

____

____

Part Three

Text B

The Smart Grid

You may have heard of the Smart Grid on the news, but not everyone knows what the grid is, let alone the Smart Grid. "The grid" refers to the electric grid, a network of transmission lines, substations, transformers and more that deliver electricity from the power plant to your home or business. Our current electric grid was built in the 1890s. To move forward, we need a new kind of electric grid to handle the groundswell of digital and computerized equipment and technology dependent on it and automate and manage the increasing complexity and needs of electricity in the 21st century.

What does a Smart Grid do?

The Smart Grid represents an unprecedented opportunity to move the energy industry into a new era of reliability, availability, and efficiency that will contribute to our economic and environmental health. The benefits associated with the Smart Grid include:

- More efficient transmission of electricity
- Quicker restoration of electricity after power disturbances
- Reduced operations and management costs for utilities, and ultimately lower power costs for consumers
- Reduced peak demand, which will also help lower electricity rates
- Increased integration of large-scale renewable energy systems
- Better integration of customer-owner power generation systems, including renewable energy systems

- Improved security

参考译文

智能电网

也许你在新闻上听说过智能电网。但并不是每个人都知道电网是什么，更不用说智能电网了。“电网”指的是由输电线路、变电站、变压器等组成的网络，将电力从发电厂输送到你的住宅或企业。我们目前的电网建于 19 世纪 90 年代。再向前发展，我们需要一种新的电网，以应对依赖于它的数字化和计算机化的设备及技术的浪潮，并且能够应对 21 世纪日益增长并复杂化的电力自动化管理需求。

智能电网是做什么的？

智能电网提供了一个前所未有的机会，使能源工业进入了一个可靠、可用和高效的新时代，这将有助于我们的经济和环境健康。智能电网的优势包括：

- 电力的高效传输
- 出现电力故障后快速修复
- 降低运营和管理成本，最终降低用户的成本
- 减少高峰需求，也将有助于降低电费
- 大规模的再生能源系统进一步融合
- 更好地整合用户发电系统，包括可再生能源系统
- 改进安全性

Words & Expressions

current ['kʌrənt]	*adj*. 现在的
groundswell ['graʊndswel]	*n*. 风潮；暴涌
digital ['dɪdʒɪtl]	*adj*. 数字的；数码的
computerized [kəm'pjuːtəˌraɪzd]	*adj*. 用计算机操作（管理）的
automate ['ɔːtəmeɪt]	*v*. 使自动化
unprecedented [ʌn'presɪdentɪd]	*adj*. 空前的；前所未有的
availability [əˌveɪlə'bɪləti]	*n*. 有效；有用；可用性
restoration [ˌrestə'reɪʃn]	*n*. 恢复；归还；复位
disturbance [dɪ'stɜːbəns]	*n*. 扰乱；骚动
peak [piːk]	*n*. 尖端；顶峰
integration [ˌɪntɪ'greɪʃn]	*n*. 集成；综合；同化
renewable [rɪ'njuːəbl]	*adj*. 可更新的；可再生的

Notes

1. the Smart Grid: 智能电网
2. renewable energy systems: 可再生能源系统

Exercises

☞ **Task 1** The following picture is a smart grid. Fill in the blanks with the right words or phrases according to the marks in the picture.

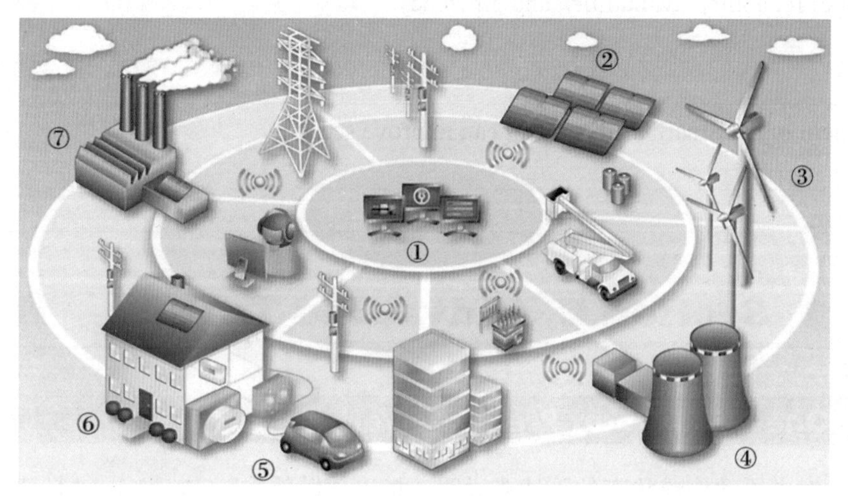

①________________

②________________

③________________

④________________

⑤________________

⑥________________

⑦________________

☞ **Task 2** Match the words with their meanings.

digital	扰乱
automate	数字的；数码的
disturbance	集成；综合
peak	使自动化
integration	尖端；顶峰

☞ **Task 3** Translate the following sentences into Chinese.

1. You may have heard of the Smart Grid on the news, but not everyone knows what the grid is, let alone the Smart Grid.

____

____

2. Our current electric grid was built in the 1890s and improved upon as technology advanced.

3. To move forward, we need a new kind of electric grid.

4. The Smart Grid represents an unprecedented opportunity to move the energy industry into a new era of reliability, availability, and efficiency.

5. One benefit of the Smart Grid is that it can improve security.

Section C Advanced Training

Part One Grammar: Predicate Verb 谓语

谓语是对主语动作或状态的陈述或说明，指出“做什么”（do what）“是什么”（what is this）或是“怎么样”（how）。谓语动词分为实义动词、情态动词、助动词和系动词四种。

谓语动词的位置一般在主语之后，以下是谓语动词的使用情况：

1. 由情态动词、助动词 + 不带 to 的动词不定式构成的复合谓语：

E.g. How does an electric grid deliver electricity to the consumers? 电网是如何把电输送给用户的？

You'd better go to the power plant on time. 你最好按时到电厂。

2. 由系动词（be, seem, feel, sound, taste, smell, become...）+ 表语（表语可以是名词、名词性短语或名词性从句、形容词、形容词性短语或介词短语）。

E.g. The three-phase system is the most usual system today. 三相系统是如今最常用的系统。

The smart grid becomes more and more popular. 智能电网越来越流行了。

3. 主语和谓语动词一致

谓语动词和主语在单、复数形式上要保持一致。寻其规律，大致可归纳为三种原则，即语法一致原则、意义一致原则和就近原则。

（1）语法一致原则

① 由连接词 and 或 both ... and 连接的合成主语，一般要用复数形式的谓语动词。

e.g. Wind power and solar power are renewable energy sources. 风能和太阳能是可再生能源。

注意：

a. 若 and 所连接的两个词是指同一个人或物时，它后面的谓语动词就要用单数形式。

e.g. The electrician and inventor is talking with my father. 那个电工兼发明人正和我父亲谈话。

b. 两个并列的名词有 each, every, many a 等修饰语时，谓语动词一般用单数。

e.g. Many a worker is trying to repair an electrical fault. 许多工人正在尝试抢修电路。

② with, as well as, along with, but, including, instead of, rather than, together with 等在主语和动词之间，是插入语，不能影响主语的单复数，谓语动词应当与主语一致。

e.g. The expert with his two assistants is in Beijing. 那个专家和他的两名助手现在在北京。

③ family, class, team, audience, army, group, public 等集体名词指整个集体时，它的谓语动词用单数；如果它指集体的成员，其谓语动词就用复数形式。

e.g. Our group has 6 members. 我们组有 6 人。

Our group are testing the alarm device.（指组内成员）

我们的组员们正在测试报警装置。

④ 不定代词 either, neither, each, somebody, someone, no one, nothing 等作主语时，谓语动词用单数。

e.g. Someone is in the control room. 有人在控制室里。

⑤ 在定语从句中，关系代词 that, who, which 等作主语时，其谓语动词的数应与先行词的数一致。

e.g. The man who is in blue is our chief engineer. 那个穿着蓝色衣服的男人是我们的总工程师。

We have already informed our staff who are on their way home. 我们已经通知那些正在回家路上的员工了。

⑥ A number of + 复数名词作主语，谓语动词用复数。

e.g. A number of our colleagues are planning to travel.（注：a number of 相当于 many）我们很多同事正计划着去旅行。

The number of + 复数名词作主语，其中心词是 number，谓语动词用单数。

e.g. The number of the employees is 206. 员工人数是 206 人。

（2）意义一致原则

① 一些学科名词是以 -ics 结尾，如：mathematics，politics，physics 以及 news，works 等，都属于形式上是复数的名词，实际意义为单数名词，它们作主语时，其谓语动词要用单数形式。

e.g. Mathematics is one of the basic courses in the college. 数学是大学基础学科之一。

② 表示“时间、重量、长度、价值”等复数名词作主语时，表示单位数量通常被看作整体，谓语动词用单数形式。

e.g. Twenty years is a long time. 20 年是一段很长的时间。

注意：如果强调数量，谓语用复数。

E.g. Three days have passed since the accident. 自从事故发生后，已经过去 3 天了。

③ 短语 / 主语从句作主语时，谓语动词用单数。

E.g. Working with my team is a joyful thing. 和我的团队一起工作是一件很开心的事情。

（3）就近原则

① here, there 开头的句子，谓语动词与最近的名词的数保持一致。

E.g. There is a wall-socket over there. 那边墙上有一个插座。

② either ... or ..., ... or ..., neither ... nor ..., not only ... but also ... 等连接两个名词作主语，谓语动词用"就近原则"。

E.g. Not only Lucy but also Max often does the research. 不仅露西，马克斯也经常做研究。

Exercises:

Choose the appropriate answer from the three choices marked A, B and C.

1. The number of the employees in Laibin thermal power plant ___________ 500.
   A. is　　B. are　　C. have
2. Two years ___________ quite a short time for my father in his career.
   A. were　　B. are　　C. is
3. Either you or your father ___________ working in the same power plant.
   A. am　　B. is　　C. are
4. The writer and engineer ___________ my uncle.
   A. has　　B. is　　C. are
5. Somebody ___________ working at the substation.
   A. has　　B. is　　C. have
6. Mike with his father ___________ visiting the hydropower station.
   A. was　　B. have　　C. is

Part Two　Writing

邀请信

邀请信包括宴会、舞会、晚餐、聚会、婚礼、活动等各种邀请信件，形式上分为两种：一种为正规的格式（formal correspondence），称为请柬或正式邀请信；请柬或正式邀请信格式严谨，邀请者用第三人称，被邀请者既可用第三人称，也可用第二人称。另一种是非正式格式（informal correspondence），即一般的邀请信。非正式的邀请信主要用于熟悉的朋友和亲人间。写信人可用第一人称提及自己，用第二人称称呼对方。邀请信的回信或复函格式与邀请信相同。

（1）邀请信的基本格式

正式邀请信的格式由写信人的地址、收信人姓名、地址、日期、称呼、正文、结束语及署名 8 个部分组成。非正式的邀请信格式可以省略写信人的地址、收信人姓名、地址和称呼。

① 信头（heading）是写信人的地址，地址书写"从小到大"，即"门牌号、街道、城市、国家"。信头一般置于信的右上角。

② 日期（Date），日期的顺序是：月、日、年，如“Jun 20, 2013”。

③ 信内地址与信头写法一样，置于左上角，日期下面。

④ 称呼：指对收信人的称呼。写在日期之下，从信纸的左边开始。一般用 Dear... 作称呼，如“Dear Jane”。在性别上，男性用 Mr.（先生）；女性有三种称呼：Miss（小姐）表示未婚女性，Mrs.（夫人）表示已婚女性，Ms.（女士）适用于所有女性。

⑤ 信的正文：指信的主体部分。一定要将邀请的时间（年、月、日、钟点）、地点、场合写清楚，不能使接信人存在任何疑虑，最后要表示期待对方接受邀请。非正式邀请信写信人可以用第一人称，用第二人称称呼对方。

⑥ 结束语：指正文下面的结尾客套话。一般从信纸的中央靠右写起，第一个字母大写，末尾用一逗号。常用 Yours Faithfully 或 Sincerely Yours。假如对方是亲密的朋友，可用 Yours 或 Love 等。

⑦ 签名：指发信人签名。写在结束语的下面，稍偏右。

正式信函的具体格式如下：

Letter head（信头）
Date（日期）
Inside Address（信内地址）
Salutation（称呼），
Body of the Letter（正文）

Yours sincerely（结束语），
Signature（签名）

非正式信件的格式如下：

Date（日期）
Salutation（称呼），
Body of the Letter（正文）

Yours sincerely（结束语），
Signature（签名）

（2）常见的例句或表达

① I would like to invite you to take part in my parents' 20th wedding anniversary at Guangming Hotel, 7 p.m. on Saturday, Dec. 21, 2018.

［译文］为庆祝我父母 20 周年结婚纪念日，定于 2018 年 12 月 21 日（星期六）晚上 7 点在光明酒店举行宴会，敬请光临。

② I am writing to invite you to visit my hometown on National Day holidays.

[译文] 我写信是邀请你国庆假期到我的家乡参观。

③ I am looking forward to seeing you.

[译文] 盼望到时能见到你。

Sample

Jun 25th, 2019

Dear June,

I am writing to invite you to come to Beijing on your summer vacation. I will show you around the Great Wall, the Imperial Palace and the Summer Palace, and make sure your stay happy and enjoyable. Please tell me whether you could come or not.

I'm looking forward to hearing from you.

Yours,

Zhang Ling

Task

说明：以 Tom 的名义给 Jack 写一封邀请信，邀请他今年五一假期到 Tom 家玩，到时带他参观他向往的企业：广西来宾 B 厂，通过参观可以熟悉电厂运营，了解设备使用情况，为以后的专业学习奠定基础。

Section D An English Song

☞ **Task** Listen to the song and fill in the blanks with the missing words you have just heard, and then sing along with it.

Save the Last Dance for Me

Michael Bublé

You can dance, every dance with the guy who gives you the eye 你可以和每个给你暗送秋波的男人跳舞

Let him hold you tight 让他紧紧拥抱你

You can smile, every smile for the man who held your hand 你可以对每个紧握你手的男人微笑

(be) neath the pale moonlight 在暗淡的月光下

But don't forget who's taking you home 但是别忘记谁带你回家

And in whose arms you're gonna be 谁的怀抱才是你的归属

So darling, save the last dance for me 哦 亲爱的，把最后一支舞留给我

Oh I know that the music's fine 哦 我知道这音乐美妙的
Like sparkling wine 就像是香醇的葡萄酒
① ______________________________

Laugh and sing 尽情欢笑，尽情歌唱
But while we're apart 不过当我们分开时
Don't give your heart to anyone 别把你的心给别人
And don't forget who's takin' you home 还有别忘了是谁带你回家
And in whose arms you're gonna be 谁的怀抱才是你的归属
So darling 哦 亲爱的
② ______________________________

Baby don't you know I love you so 宝贝难道你不知道我是多么爱你
Can't you feel it when we touch 当我们彼此触碰时难道你感觉不到吗
I will never, never let you go 我将永远不会让你离去
I love you oh so much.（因为）我是这样的爱你
You can dance, go and carry on 你可以跳舞，跳啊跳
till the night is gone 一直从长夜跳到黎明
And it's time to go 跳到曲终人散时
If he asks if you're all alone 然后如果他问你是否独自一人
③ ______________________________

you must tell him no 你一定要告诉他“不”
Cause don't forget who's taking you home 因为别忘记是谁带你回家
And in whose arms you're gonna be 谁的怀抱才是你的归属
save the last dance for me 把最后一支舞留给我
Oh I know that the music's fine 哦 我知道这音乐美妙的
Like sparkling wine, Go and have your fun 就像是香醇的葡萄酒，去尽情玩儿吧
Laugh and sing 尽情欢笑，尽情歌唱
But while we're apart 不过当我们分开时
Don't give your heart to anyone 别把你的心给别人
And don't forget who's taking you home 还有别忘了是谁带你回家
And in whose arms you're gonna be 谁的怀抱才是你的归属
So darling, save the last dance for me 哦 亲爱的，把最后一支舞留给我

So don't forget 因为别忘记

④ ______________________________

And in whose arms you're gonna be 谁的怀抱才是你的归属

So darling, save the last dance for me 哦 亲爱的把最后一支舞留给我

Oh baby won't you save the last dance for me 哦，宝贝，把最后一支舞留给我吧

Background Tip:

Michael Bublé。麦可·布雷，加拿大著名流行爵士乐歌手，也是一位影视演员。他 2003 年发表的专辑《麦克·布雷》(*Michael Bublé*)、2004 年的《与我共遨游》(*Come Fly With Me*) 以及 2005 年的《是时候了》(*It's Time*) 让他成功地在加拿大、英国、澳大利亚以及美国打响了名堂。*Save the Last Dance for Me* 是他的翻唱作品(原唱 The Drifters)。

Unit 3

Power Plant Equipment

电厂设备

Section A Professional Background

Part One Lead-in

☞ **Task** Study the pictures and discuss the questions below.

Questions:

1. What kind of power station is shown in the first picture?
2. What can you see in the first picture?
3. What is the equipment in the second picture?

Cues:
tipping hall, grab-crane, control room, scrubber, bag-house filter, steam, turbine

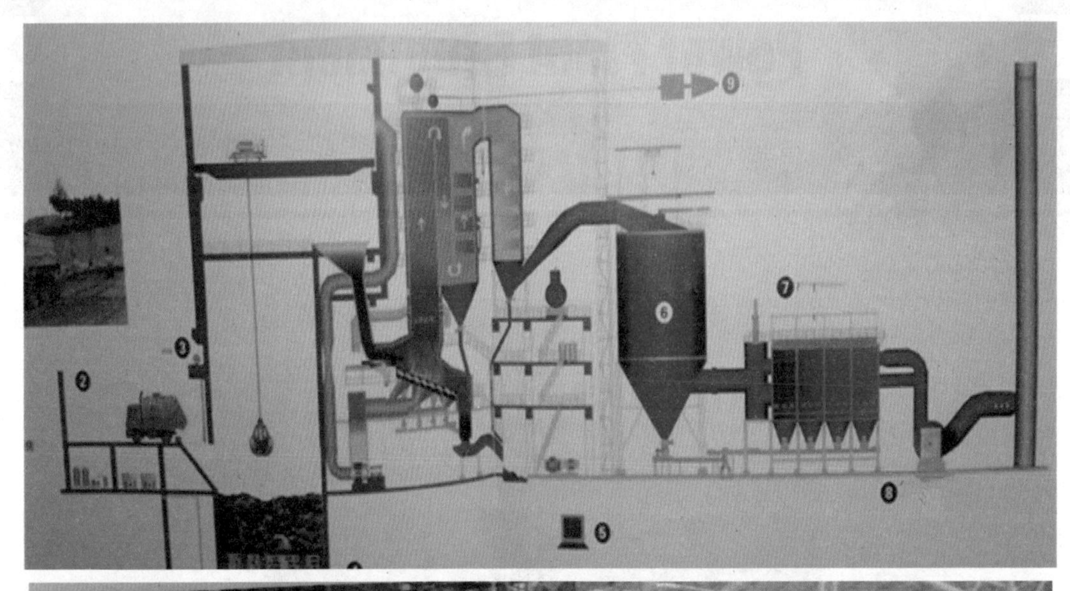

Part Two Listening & Speaking

Listen to the following dialogues and fill in the blanks.

Dialogue I

W: Hey Jack, have you got any plans for the coming May Day __________?

M: Tom __________ me to his home.

W: Why his home? Is there anything ___________?

M: You know I major in Thermal Energy Dynamic Installation of Power Plant. (电厂热能动力装置).

W: What does that matter?

M: Tom's ___________ lives in the Guangxi Laibin Power Plant A.

W: So you want to visit the power plant?

M: Yes, Tom heard I am ___________ to visit the power plant for a long time.

W: Oh, I see.

☞ Dialogue 2

W: Jack, you have fulfilled your ___________ of visiting a power plant. How do you like it?

M: Tom ___________ me around the whole plant. I love the environment.

W: Do you mean the living environment or the ___________ environment?

M: Both.

W: Really? Tell me more.

M: More than expected, I saw the ___________ equipment like boiler, generator, control room and so on.

W: So you have prepared for your practice.

M: Yes. I have an interesting ___________.

Section B　Text Learning

Part One　Lead-in

1. Do you know the components of a nuclear power plant?
2. What is the function of a generator?

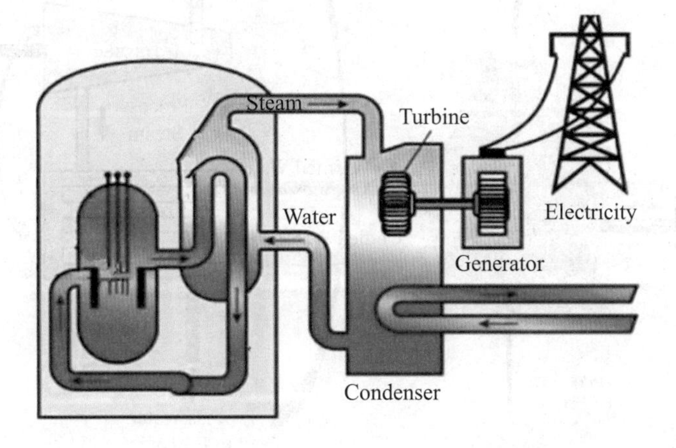

Part Two

Text A

Basic Components of a Power Plant

The basic components of an electrical power plant may be shown as in Fig. 3.1. In the figure, the boiler, turbine, condenser and the generator are shown in the middle, and the transformer, switching station and the transmission line are shown on the right.

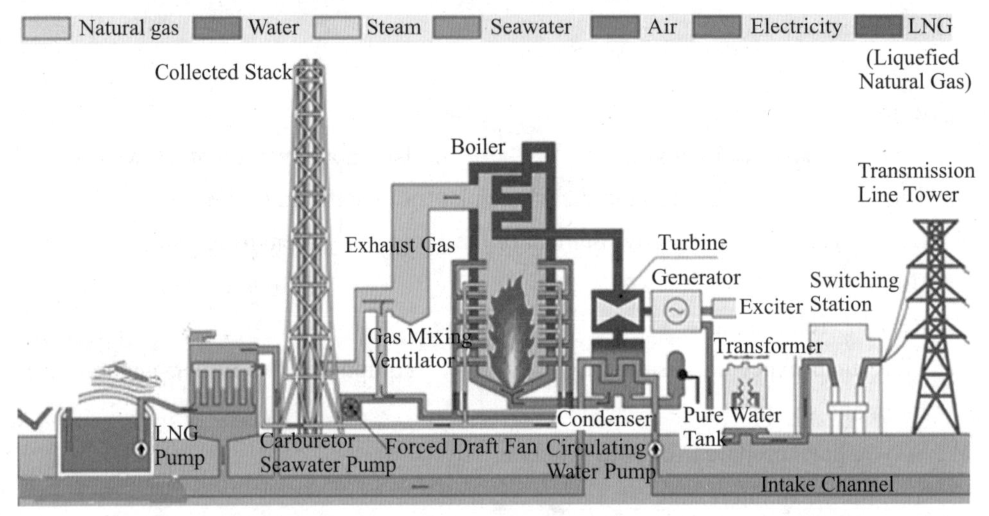

Fig. 3.1

Boiler is a device that transfers the heat from the products of combustion (coal, oil, or natural gas) to water and produces hot water or steam.

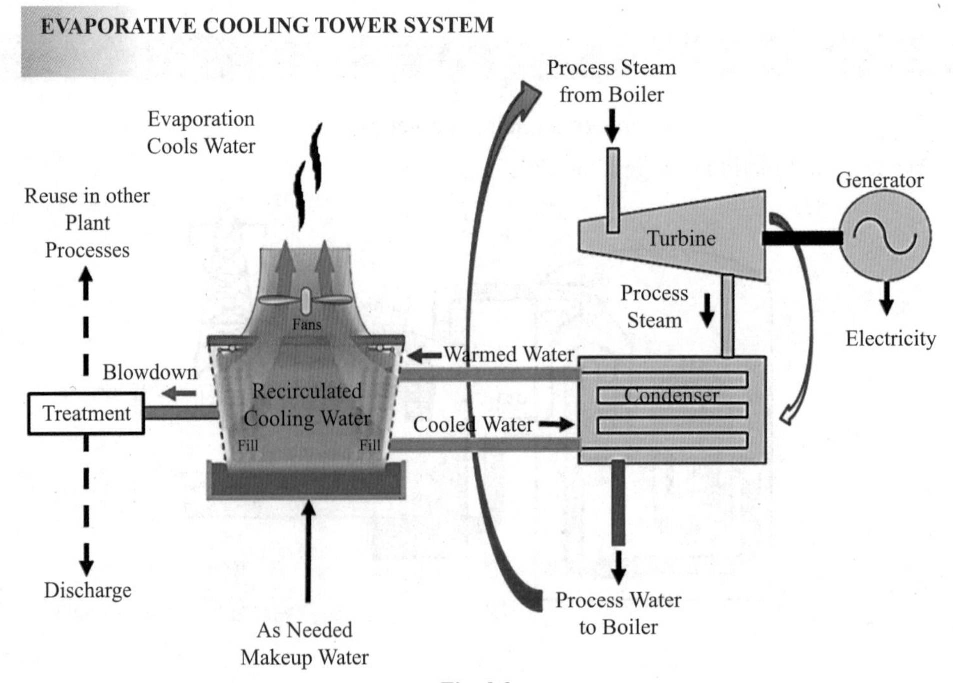

Fig. 3.2

A steam condenser (Fig. 3.2) is a vessel in which exhaust steam is condensed by contact with cooling water. Condenser condenses the steam back to water so that it can be returned to the heat source to be heated again.

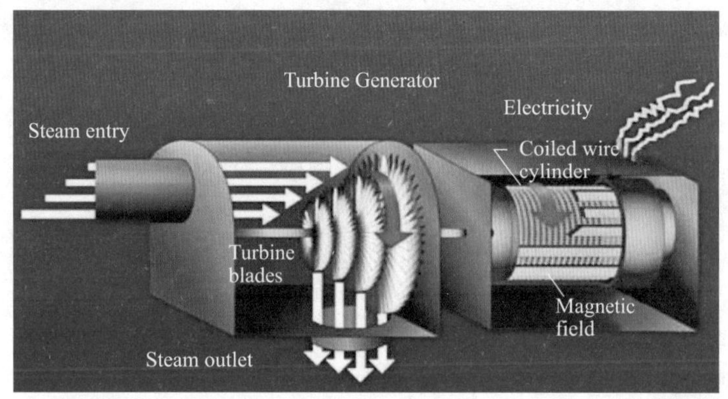

Fig. 3.3

Turbine (Fig. 3.3) converts the kinetic energy of the steam to the mechanical energy.

Electric generator (Fig. 3.4) is coupled with the turbine rotor and converts the mechanical energy of the turbine to the electrical energy.

A transformer (Fig. 3.6) is an electrical device that transfers electrical energy between two or more circuits through electromagnetic induction. Over long distance transmission, high voltage is required. Electric transformer can change the normally low voltage to the high voltage needed for the efficient transmission. Transformer can only work on alternating current (AC).

Pump (Fig. 3.1) provides the force to circulate the water through the system.

Fig. 3.6

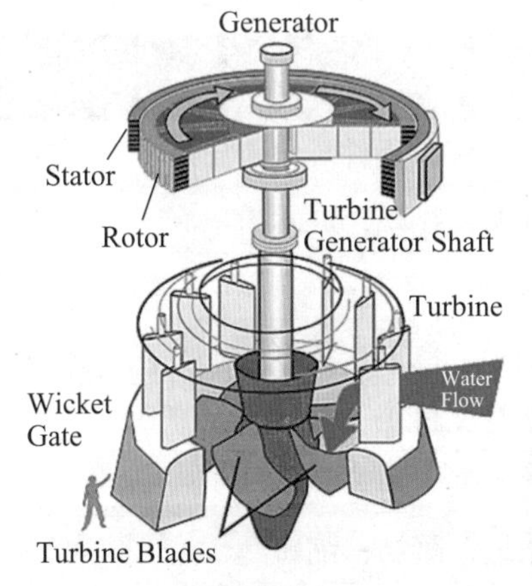

Fig. 3.4

Fig. 3.5

Power plants boil water to create steam, which then rotates turbines to generate electricity. Once steam has passed through a turbine, it must be cooled and condensed back into liquid water to start the cycle over again to produce more electricity. Cooling water cools the steam more effectively and allows more efficient electricity generation. Power plants typically have cooling towers (Fig. 3.5) to reduce the amount of water obtained from the nearby lake, river, or ocean and also reduce the amount of thermal pollution discharged from the plants back into those bodies of water.

(297 words)

参考译文

发电厂的基本部件

发电厂的基本组成部分如图 3.1 所示。在图中，中间是锅炉、涡轮、冷凝器和发电机，右侧是变压器、开关站和传输线。

锅炉是一种将燃烧物（煤、油或天然气）的热量传送到水中产生热水或蒸汽的设备。

蒸汽冷凝器（图 3.2）是一种通过与冷却水接触而凝结废气的容器。冷凝器把蒸汽冷凝成水，这样就可以把它送回热源再加热。

涡轮机（图 3.3）将蒸汽的动能转换成机械能。

发电机（图 3.4）与涡轮转子耦合，将涡轮的机械能转换为电能。

变压器是一种电气设备，通过电磁感应在两个或多个电路之间传输电能。长距离传输的话就需要高电压。电力变压器可以把低电压变成高效率传输所需的高电压。变压器只能传输交流电。

泵（图 3.1）提供水在循环系统内流动所需的动力。

发电厂将水烧开产生蒸汽，然后旋转推动涡轮机发电。蒸汽一旦通过涡轮，就必须冷却并冷凝回液态水，才能重新开始循环来产生更多的电力。冷却水能更有效地冷却蒸汽，从而提高发电效率。发电厂往往有冷却塔（图 3.5），以减少从附近湖泊、河流或海洋中获取的水量，并减少从发电厂排放回水体的热污染。

Words & Expressions	
component [kəm'pəunənt]	*n.* 零部件；元件
show [ʃəu]	*v.* 证明；显现；展示
combustion [kəm'bʌstʃən]	*n.* 燃烧
boiler ['bɔɪlə(r)]	*n.* 锅炉；汽锅；热水器
turbine ['tɜːbaɪn]	*n.* 涡轮（机）
condenser [kən'densə]	*n.* 冷凝器；电容器
condense [kən'dens]	*v.* 浓缩；凝结；缩短
vessel ['vesl]	*n.* 容器；器皿
exhaust [ɪg'zɔːst]	*v.* 耗尽 *n.* 排气装置；废气
kinetic [ki'netɪk]	*adj.* 运动的；动力学的

mechanical [mə'kænɪkl]	*adj*. 机械的；力学的
circuits ['sɜːkɪts]	*n*. 电路 circuit 的复数形式
coupled ['kʌpld]	*adj*. 成对的（连接的；共轭的；联系的）
rotor ['rəʊtə(r)]	*n*. 旋转体；[机] 转子
normally ['nɔːməli]	*adv*. 通常；正常地
electromagnetic [ɪˌlektrəʊmæg'netɪk]	*adj*. 电磁的
induction [ɪn'dʌkʃn]	*n*. 感应；诱发
rotate [rəʊ'teɪt]	*v*. 旋转；循环

Notes

1. kinetic energy: [力] 动能
   e.g. We can find that the atoms' kinetic energy also increases.
   我们能发现原子的运动能量也增加了。
   kinetic hydraulic energy: 水力动能
2. mechanical energy: 机械能机器能够将输入的各种能量转变为机械能 (Mechanical energy)，例如刨床将输入的电能转变为机械能而做功。
   mechanical heat energy 机械热能
3. be coupled with: 与……结合；与……连接；与……配合
   e.g. In the multimedia classroom, the monitor is coupled with the mainframe.
   在多媒体教室里，显示器与主机相连。
4. convert to: 转换至
   e.g. We need to convert from flv to other format.
   我们需要将视频格式转换成其他格式。

Exercises

☞ **Task 1**　Answer the following questions according to the text.

1. Can you tell me how many basic components of an electric power plant may be shown in Fig. 3.1?

   ____

   ____

2. What is the function of a boiler?

   ____

   ____

3. What can the turbine convert the kinetic energy of the steam to?

   ____

   ____

4. What can an electric generator convert the mechanical energy of the turbine to?

____

____

5. What kinds of device can change the normally low voltage to the high voltage needed for the efficient transmission?

____

____

☞ **Task 2** Change the word formation.

Before	After	Meaning
boiler *n.*		*v.* 沸腾；煮沸
condenser *n.*		*v.* 浓缩；凝结成液体
coupled *adj.*		*n.* 对；夫妇；数个
combustion *n.*		*adj.* 可燃的
rotate *v.*		*n.* 旋转；循环；轮流

☞ **Task3** Complete the following sentences with the proper form of the words from Task 2.

1. Steam can be ___________ into water.
2. Paper is a ___________ object.
3. ___________ is one of the most important equipment in the thermal power plant.
4. They are a nice ___________.
5. High steam pushes the turbine blades to ___________.

☞ **Task 4** Translate the following sentences into Chinese.

1. Boiler is a device that transfers the heat from the products of combustion (coal, oil, or natural gas) to water.

____

____

2. Turbine converts the kinetic energy of the steam into mechanical energy.

____

____

3. A transformer is an electrical device that transfers electrical energy between two or more circuits through electromagnetic induction.

____

____

4. Transformer can only work on alternating current.

____

____

5. Cooling water can be obtained from a nearby river or lake.

____

____

Part Three

Text B

Why Do We Use High Voltages?

If our offices and homes using computers, photocopiers, washing machines and electric shavers rated at 110-250 volts, why don't power station simply transmit electricity at that voltage? Why do they use such high voltages? To explain that, we need to know a little about how electricity travels.

As electricity flows down a metal wire, the electrons carry its energy through the mental structure, bashing and crashing about and wasting energy like naughty children running down a corridor. That's why wires get hot when electricity flow through them. It turns out that the higher the voltage you use, and the lower the current, the less energy is wasted. So the electricity that comes from power plants is sent down at extremely high voltages to save energy.

But there's another reason too. Factories have huge machines that are much bigger and more energy-hungry than you have at home. The energy an appliance uses is directly related (proportional) to the voltage it uses. Power-hungry machines might use 10,000-30,000 volts. It makes sense to carry high-voltage electricity from the power station and then transform it to lower voltages when it reaches its various destinations.

(179 words)

参考译文

我们为什么要使用高压？

如果我们的办公室和家庭使用的是110—250伏的电脑、复印机、洗衣机和电动剃须刀，为什么发电站不直接在这个电压下传输电呢？他们为什么要用那么高的电压？为了解释这一点，我们需要了解一下电力是如何运行的。

当电流沿着金属电线流动时，电子携带着能量在金属结构中四处撞击，浪费能量，就像顽皮的孩子在走廊里奔跑一样。所以，当电流通过时电线会变热。事实上，你用的电压越高，电流越低，浪费的能量越少。所以，来自发电厂的电才会以极高的电压传输以节省能源。

但还有另一个原因。工厂有巨大的机器，它们比你家里的任何机器都要大得多，也更耗电。电器所使用的能量与它所使用的电压成正比。耗电机器可能会使用高达1万伏到3万伏的电压。从发电站运送高压电力，然后在到达不同的目的地时将其转换成较低的电压，这是有意义的。

Words and Expressions

Words and Expressions	
photocopier ['fəʊtəʊkɒpɪə(r)]	*n.* 复印机；影印机
shaver ['ʃeɪvə(r)]	*n.* 剃须刀
flow [fləʊ]	*v.*（使）流动；传播
metal ['metl]	*n.* 金属；合金 *adj.* 金属制的
electron [ɪ'lektrɒn]	*n.* 电子
bash [bæʃ]	*v.* 猛击，痛击；严厉批评
crash [kræʃ]	*v.* 碰撞；（使）摔碎；（机器、系统等）崩溃
corridor ['kɒrɪdɔː(r)]	*n.* 走廊
current ['kʌrənt]	*n.*（水、气、电）流；趋势；涌流 *adj.* 现在的；流通的，最近的
extremely [ɪk'striːmli]	*adv.* 非常，极其；极端地
appliance [ə'plaɪəns]	*n.* 器具；器械；装置；应用
destination [ˌdestɪ'neɪʃn]	*n.* 目的地，终点
turn out	发生；结果是；出席；熄灭；生产，制造
make sense	有意义；讲得通；言之有理

Notes

1. It turns out that...: 结果是……，证明是……，原来……
   e.g. It turns out that if you change how people talk, that changes how they think.
   事实上，如果你改变人们说话的方式，就会改变他们思考的方式。
2. Power-hungry machines might use 10, 000—30, 000 volts.
   耗电机器可能会使用高达 1 万伏到 3 万伏的电压。（较小的工厂和机械车间需要 400 伏左右的电压。）
3. make sense: 有意义；讲得通；言之有理
   e.g. It doesn't make sense. 这不合常理 / 这没有意义。

Exercises

☞ **Task 1** Write down the words according to their explanations.

1. ____________	*n.* a machine that makes copies of documents
2. ____________	*v.* to move steadily and continuously in one direction
3. ____________	*n.* a hard substance such as iron, gold, or steel
4. ____________	*n.* a flow of electricity through a wire
5. ____________	*adv.* to a very high degree

☞ **Task 2**　Match the following items according to the pictures.

A.　　B.　　C.　　D.　　E.

1. photocopier　　2. shaver　　3. metal　　4. crash　　5. appliance

☞ **Task 3**　Translate the follow sentences into English.

1. 他们为什么要用那么高的电压？

____

____

2. 当电流通过时电线会变热。

____

____

3. 你用的电压越高，电流越低，浪费的能量越少。

____

____

4. 电器所使用的能量与它所使用的电压成正比。

____

____

5. 耗电机器可能会使用高达 1 万伏到 3 万伏的电压。

____

____

Section C　Advanced Training

Part One　Grammar: Basic Sentence Pattern 英语基本句型

英语中的五种基本句型结构有：

句型一：Subject（主语）+ Verb（谓语）

本句型的谓语动词大多是不及物动词，常见的不及物动词如：rain，happen，leave，work 等。例如：

（1）My father works in a nuclear power plant. 我父亲在一家核电厂工作。

（2）It rained last night. 昨晚下雨了。

句型二：Subject（主语）+ Verb（谓语）+ Object（宾语）

这种句型中的谓语动词一般为及物动词，及物动词后需接宾语，不接宾语句子表达不完整。其宾语通常由名词、代词、动词不定式、动名词或从句等来充当。

例如：

（1）He is boiling water. 他正在烧水。（完整）

（2）Power plants have... 电厂有……（不完整，有什么）

改为：Power plants have boiler, turbine, generator and other equipment.

电厂有锅炉、涡轮机、发电机及其他设备。

句型三：Subject（主语）+ Link Verb（系动词）+ Predicate（表语）.

这种句型主要用来表示主语的特点、身份等。其系动词一般可分为下列两类：

1. 表示状态。

这样的词有：be, look, seem, smell, taste, sound, keep 等。可以用形容词、名词短语或介词短语作表语。例如：

（1）A transformer is an electrical device. 变压器是一种电气设备。

（名词短语作表语）

（2）The smoke from chimney smells bad. 烟囱的烟闻起来很难闻。

（形容词作表语）

（3）Tom is at the foot of a cooling tower. 汤姆在冷却塔下面。

（介词短语作表语）

2. 表示变化。这类系动词有：become, turn, get, grow, go 等。

例如：Steam becomes cold in the condenser. 蒸汽在冷凝器中变冷。

句型四：Subject（主语）+ Verb（谓语）+ Indirect Object（间接宾语）+ Direct Object（直接宾语）

这种句型中，多由指“物”的名词承担直接宾语，表示动作是对谁做的或为谁做的，在句中不可或缺，间接宾语通常由代词来承担。引导这类双宾语的常见动词有：buy, pass, lend, give, tell, teach, send 等。例如：

（1）My mother gave me a thermos last night. 妈妈昨晚给了我一个热水瓶。

（2）I pass him the magnet. 我把磁铁递给他。

句型五：Subject（主语）+ Verb（动词）+ Object（宾语）+ Complement（补语）

这种句型中的“宾语+补语”统称为“复合宾语”。宾语补足语的主要作用或者是补充、说明宾语的特点、身份等，或者表示让宾语去完成的动作等。担任补语的常常是名词、形容词、副词、介词短语、分词、动词不定式等。例如：

（1）We should keep our power station clean and tidy. 我们应该保持电厂的干净整洁。

（形容词）

（2）They made Tom our director. 他们让汤姆做我们的主管。

（名词）

（3）I told you not to do that in the control room. 我告诉你不要在控制室那么做。
（不定式）

（4）I found my car missing near the waste-to-energy plant. 我发现我的车在垃圾焚烧发电厂附近不见了。　（现在分词）

常见跟补语的动词有：tell, ask, advise, help, want, would like, order, force, allow 等。

注意：动词 have, make, let, see, hear, notice, feel, watch 等后面所接的动词不定式作宾补时，不带 to。例如：

（1）He made me take the picture by the equipment. 他让我在设备旁照相。

（2）I heard her play the piano last night. 我昨晚听到她弹钢琴。

Exercises

Choose the appropriate answer from the three choices marked A, B and C.

1. I feel ___________ in traveling in the biggest solar power station.
   A. interested　　B. interest　　C. to interest
2. My mother is ___________.
   A. on the top tower　　B. top tower　　C. of the tower
3. My father ___________ in Laibin Thermal Power Station for 25 years.
   A. works　　B. has worked　　C. would work
4. I like ___________ in the library.
   A. reading　　B. read　　C. is reading
5. The boss had me ___________ for a long time in the plant.
   A. to wait　　B. waits　　C. wait
6. I told you ___________ near the high voltage tower.
   A. not to play　　B. plays　　C. will play

Part Two　Writing

感谢信

感谢信（Letter of Thanks）是机构或个人对关心、支持、帮助或热情款待表示感谢的对外函件。其具体格式和要求与邀请函相同。写作时须说明感谢的内容，表述时要目标明确，语言贴切、充分、言之有物，表达出感激之情。

常用词句：

1. I am writing to express my thanks/ gratitude/ appreciation for your assistance.
   我写信是为了感谢你的帮助。
2. With your help, I have good cooperation with other members during work.
   在你的帮助下，我在工作中才能和其他成员合作顺利。
3. Without your assistance and care, I can not make good performance definitely.

没有你的帮助和照顾，我肯定没有这么好的表现。

4. In a word, my appreciation to you is beyond words.
   总之，我对你的感激尽在不言中。
5. In the years to come, I will make more effort to repay your kindness.
   在接下来的几年里，我会加倍努力回报你的好意。

Sample

Jun 20, 2018

Dear Mr. Winston,

Thank you very much for the interview yesterday. I learned a great deal of your company such as main products, current developing pattern and your future blueprint in Nanning.

I believe I am fully qualified for the work you described. My internship experience is directly related to the work you are offering. Besides, my academic background and the training I received provide a strong base for my further career development.

Thanks again for the interview. I look forward to hearing from you soon.

Sincerely yours,
Li Xiaoming

☞ **Task**

说明：假设你叫 Tom，请你给部门主管 Mr. William 写一封感谢信，感谢他在你实习期间带你了解电厂运营及设备情况，向你传授工作经验，帮助你制定职业发展规划，提高了你的专业能力。此外，请你表明实习结束后你愿意继续留在电厂做一名正式员工。（注意书信的格式）

Section D　An English Song

☞ **Task**　Listen to the song and fill in the blanks with the missing words you have just heard, and then sing along with it.

Dream On

Aerosmith

Every time that I look in the mirror 每一次，看着镜子里
All these lines on my face gettin' clearer 脸上的皱纹越来越清晰
The past has gone 过去已逝
It went by like dusk to dawn 就像黄昏到黎明
Isn't that the way 人生亦如此
Everybody's got their dues in life to pay 人生总要有付出
Yeah, I know nobody knows where it comes and where it goes
是，我也知道，没有人清楚何去何从
I know it's everybody's sin 我知道，这是大家的罪过
You got to lose to know how to win 你一定要先失败，才知道怎么成功
Half my life's in book's written pages 我的人生之书已完成一半
Lived and learned from fools and from sages 从贤者与愚者身上吸取
You know it's true 你知道，这是事实
① ____________________

Sing with me, sing for the year 跟我一起唱，为了岁月
Sing for the laughter and sing for the tear 为了欢笑，为了眼泪
Sing with me, if it's just for today 跟我为当下而唱
Maybe tomorrow someone will take you away 因为说不定明天你就会被带走
Sing with me, sing for the year 跟我一起唱，为了岁月
Sing for the laughter and sing for the tear 为了欢笑，为了眼泪
Sing with me, if it's just for today 跟我为当下而唱
Maybe tomorrow someone will take you away 因为说不定明天你就会被带走
Dream on, dream on, dream on 继续梦想，继续梦想，不断梦想
② ____________________

Dream on, dream on, dream on 继续梦想，继续梦想，不断梦想
And dream until your dream comes true 梦想，直到你梦想成真
Dream on, dream on, dream on, dream on 追梦，追梦，追梦，追梦
Dream on, dream on, dream on ... aaahhh 追梦，追梦，追梦……
Sing with me, sing for the year 跟我一起唱，为了岁月
Sing for the laughter and sing for the tear 为了欢笑，为了眼泪
Sing with me, if it's just for today 跟我为当下而唱

Maybe tomorrow somconc will takc you away 因为说不定明天你就会被带走
Sing with me, sing for the year 跟我一起唱，为了岁月
Sing for the laughter and sing for the tear 为了欢笑，为了眼泪
Sing with me, if it's just for today 跟我为当下而唱
Maybe tomorrow someone will take you away 因为说不定明天你就会被带走
③ ____

Sing for the laughter and sing for the tear 为了欢笑，为了眼泪
Sing with me, if it's just for today 跟我为当下而唱
Maybe tomorrow someone will take you away 因为说不定明天你就会被带走
Sing with me, sing for the year 跟我一起唱，为了岁月
Sing for the laughter and sing for the tear 为了欢笑，为了眼泪
④ ____

Maybe tomorrow someone will take you away
因为说不定明天你就会被带走

Background Tip:

空中铁匠乐队（Aerosmith），是 20 世纪 70 年代最受欢迎的摇滚乐队之一。他们常常被称为“来自波士顿的坏男孩”和“美国最伟大的摇滚乐队”。其作品风格根植于布鲁斯的硬摇滚，同时融合了流行音乐和重金属元素。该乐队曾 21 次进入《公告牌百强单曲榜》的前 40 名，9 次排主流摇滚排行榜第一名，4 次获得格莱美大奖，10 次获音乐电视奖。在 2001 年的时候，Aerosmith 进入了摇滚名人堂。乐队的成名之路是饱经风霜和挫折的，而最终成功的秘诀是他们一直没有放弃过追求自己的梦想。这首 *Dream On* 也是他们最早的一首成名曲，收录在专辑 *Aerosmith* 中。

Unit 4

Thermal Power Plant

火电厂

Section A Professional Background

Part One Lead-in

☞ **Task** Study the pictures and discuss the questions below.

Questions:

1. What kind of power station can you see in the first picture?
2. What is the function of a transformer?
3. What can you see in the last two pictures?

Cues:

power station, power plant, transformer, kilovolt(kv), control room, turbine, cooling tower, chimney

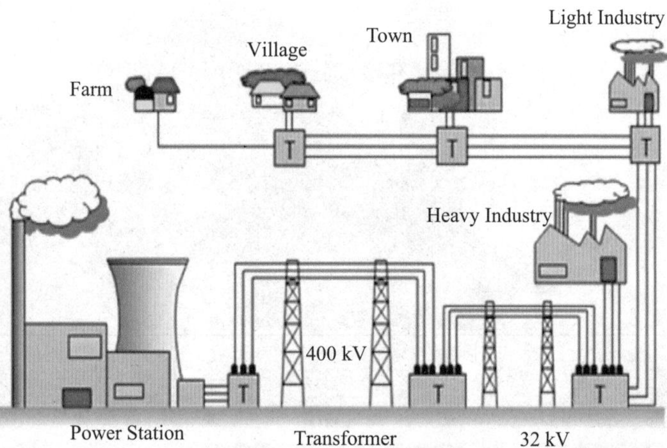

Part Two　Listening & Speaking

Listen to the following dialogues and fill in the blanks.

Dialogue 1

W: Tom, you are a ___________. Do you have any plans this term?

M: Yes, after ___________ the power plant, I know what I should do in college.

W: What is your plan in college?

M: As I have known, a graduate is not only ___________ at major but also at ___________ quality.

W: How to enhance your quality?

M: I can join in some associations in college to enhance my ability.

W: What abilities do you think will a company require the graduates to have?

M: I think they require professional ability, analysis ability, information processing, communication skills and ___________ skills, English ability and so on.

W: I want to learn from you.

M: OK.

Dialogue 2

W: I ___________ many associations will recruit ___________ community next week. Which one do you want to join in?

M: The Electrician Association.

W: Why?

M: I want to practice my professional ability. What about you?

W: I want to join in the Speech and English Associations.

M: Why?

W: Because I want to practice my English ability, communication skills and interpersonal skills.

M: I think you will have a ___________ college life.

W: Thank you. You must learn a lot in the Electrician Assoclation.

M: Yes, in this association, I will learn to ___________ other members and we can ___________ the professional problems.

Section B Text Learning

Part One Lead-in

1. What is the function of a boiler in a thermal power plant?
2. How does a thermal power plant work?
3. What is the most important index (指标) of a thermal power plant?

Part Two

☞ **Text A**

Thermal Power Plant

Thermal power plant is also referred to as coal thermal power plant or steam turbine power plant. The various components of the thermal power plant are:

Coal storage is the place where coal is stored before required.

Coal handling is designed to convert coal into the pulverized form before feeding to the boiler furnace.

Ash storage is to store ash from the burnt coal.

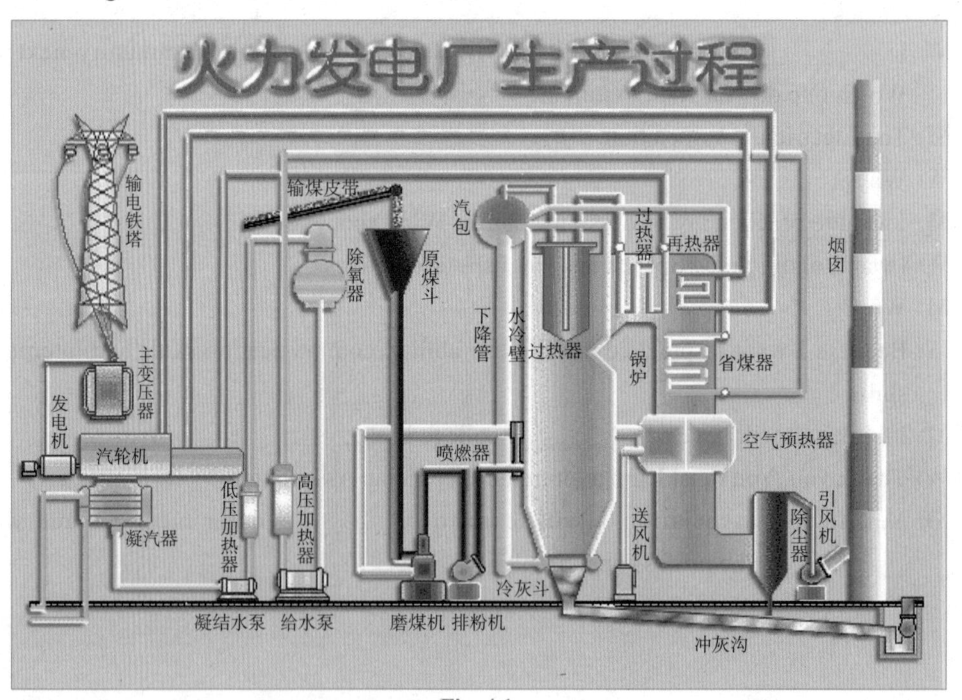

Fig. 4.1

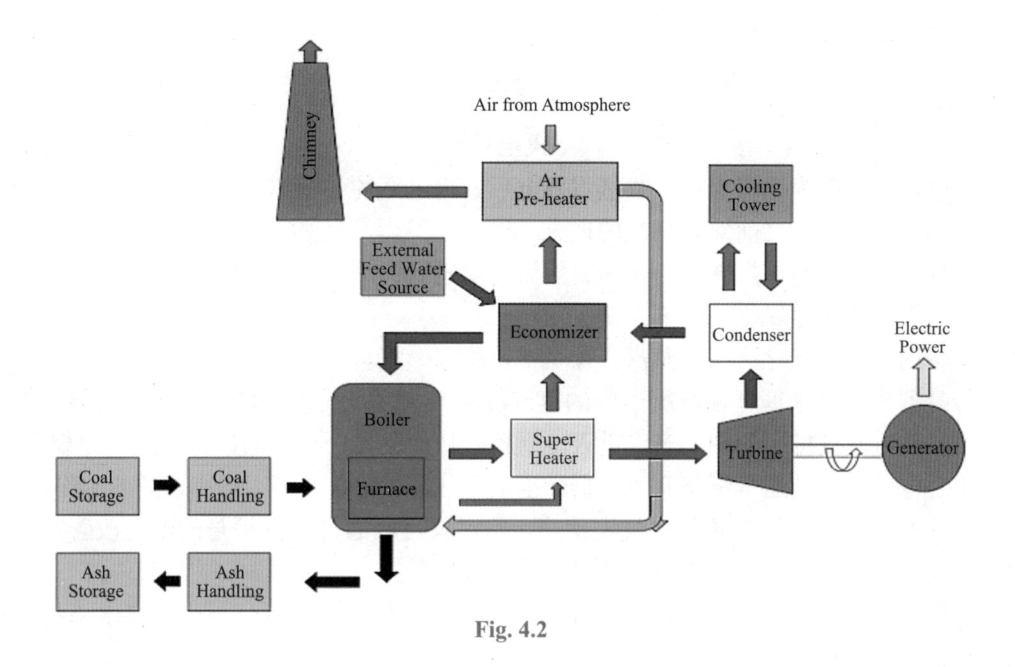

Fig. 4.2

Boiler is used to convert the water into high pressure steam. The combustion of coal takes place in the furnace.

Condenser condenses the exhausted steam in the condenser by means of cold water circulation. It converts the low pressure steam to water. It is attached to the cooling tower. Cooling tower contains cold water from the condenser for the cooling of the residual steam from the turbine.

Central station boilers are usually equipped with an economizer and an air pre-heater. Economizer can economize the working of the boiler. Air-preheater, which not only improves boiler by lowering the stack temperature but also improves the combustion condition by raising the combustion air temperature, is used to pre-heat the air before entering into the boiler furnace.

Chimney is used to release the hot burnt gases or smoke from the furnace to the environment at appropriate height. Feed water pump is used to transport the feed water to the boiler.

Turbine is the mechanical device which converts the kinetic energy of the steam to the mechanical energy. High pressure super heated steam is fed to the steam turbine to push turbine blades to rotate.

Generator is coupled with the turbine rotor and converts the mechanical energy of the turbine to the electrical energy. The electricity generated through the generator coupled to the turbine is then fed to the main grid via a system of transformers and other electrical equipment.

(300 words)

参考译文

火电厂

火电厂也被称为燃煤火电厂或蒸汽发电厂。火电厂的各个组成部分是：

煤场是在需要煤之前储存煤炭的地方。

处理煤的目的是将煤转化为煤粉，然后再输送给锅炉炉膛。

贮灰场用于储存煤燃烧后的煤灰。

锅炉用于将水转化为高压蒸汽。煤是在炉膛里燃烧。

冷凝器通过冷水循环使冷凝器内的蒸汽冷凝。它将低压蒸汽转化为水。它与冷却塔相连接。从汽轮机排出的剩余蒸汽进入冷凝器，从冷凝器中出来的冷却水进入冷却塔。

中心电站的锅炉通常配备有省煤器和空气预热器，省煤器可以提高锅炉的工作效率。空气预热器将进入锅炉炉膛的空气先预热。它不仅可以通过降低炉身的温度改善锅炉的效率，而且可以通过提高燃烧空气的温度改进燃烧条件。

烟囱用于将熔炉中高温的燃烧气体或烟以适当的高度释放到环境中。供水泵将供水输送给锅炉。

涡轮机是将蒸汽的动能转换成机械能的机械装置。高压过热蒸汽被送入汽轮机推动涡轮叶片旋转。

发电机 / 交流发电机与涡轮转子耦合，将涡轮的机械能转换为电能。通过发电机和涡轮产生的电力通过变压器和其他电气设备进入主电网。

Words & Expressions

required [rɪ'kwaɪəd]	*adj.* 需要的
handling ['hændlɪŋ]	*n.* 处理；操作
design [dɪ'zaɪn]	*n.* 设计；图样 *v.* 设计
pulverized ['pʌlvəraɪzd]	*adj.* 毁坏的（动词 pulverize 的过去式及过去分词形式）
feed [fiːd]	*v.* 供给，提供 *n.* 饲料；饲养
storage ['stɔːrɪdʒ]	*n.* 贮藏；仓库
combustion [kəm'bʌstʃən]	*n.* 燃烧
furnace ['fɜːnɪs]	*n.* 炉子；炉膛
circulation [ˌsɜːkjə'leɪʃn]	*n.* 流通；循环
attach [ə'tætʃ]	*v.* 附上；系上；贴上
residual [rɪ'zɪdjʊəl]	*adj.* 剩余的；残余的 *n.* 剩余部分

equip [ɪ'kwɪp]	*vt.* 装备；具备 *n.*（*abbr.*）装备（=equipment）
economizer [ɪ'kɒnəˌmaɪzə]	*n.* 省煤器
air-preheater [ˌeərpriː'hiːtə]	*n.* 空气预热器
stack [stæk]	*n.*（一）堆，（一）叠；烟囱；许多；排气管 *v.* 堆积；堆放
release [rɪ'liːs]	*n.* 释放；发行 *vt.* 释放
kinetic [kɪ'netɪk]	*adj.* 运动的
appropriate [ə'prəuprɪət]	*adj.* 适当的；相称的
height [haɪt]	*n.* 高度；高处；顶点
kinetic energy	动能
mechanical energy	机械能

Notes

1. feed to: 供应给
   e.g. The herdsman has to feed grass to cows every day.
   牧民必须每天给奶牛喂草料。
2. take place: 发生
   e.g. Their argument took place in the dorm last weekend.
   上周末他们在宿舍里发生了争吵。
3. by means of: 借助于，用，依靠
   e.g. Girls become more beautiful in the photos by means of PS technology.
   借助 PS 技术，照片里的女孩们变得更漂亮了。
4. be attached to: 喜爱；附属于，被连接在
   e.g. Tag is attached to the computer.
   把标签贴在电脑上。
5. be equipped with: 配备有……；装有……；装备有……
   e.g. Multimedia classrooms are always equipped with projectors.
   多媒体教室总是配备有投影仪。
6. via: *prep.* 取道，通过；经由
   e.g. He flew to Beijing via Nanning.
   他取道南宁飞向北京。

Exercises

☞ **Task I** Answer the following questions according to the text.

1. What is coal storage used for?

____

2. What is coal handling designed to?

____

3. What is feed water pump used to?

____

4. Which two kinds of device are central station boilers usually equipped with?

____

5. Why is high pressure super heated steam fed to the steam turbine?

____

☞ **Task 2** Change the word formation.

Before	After	Meaning
handling *n.*		*vt.* 处理；对待；操作
pulverized *adj.*		*vt.* 磨成粉；粉碎
kinetic *adj.*		*adv.* 运动地
residual *n.*		*adv.* 残余地；剩下地
storage *n.*		*vt.* 储存；贮藏；保存

☞ **Task 3** Complete the following sentences with the proper form of the words from Task 2.

1. Hammer mill (磨煤机) is used to ___________ coal before entering into the boiler.
2. ___________ steam from the turbine is condensed by condenser into cool water.
3. The electricity voltage must be ___________ during the transmission and distribution system.
4. Coal can be ___________ in the coal storage.
5. Turbine can convert mechanical energy into ___________ energy.

☞ **Task 4** Translate the following sentences into Chinese.

1. Coal storage is the place where coal is stored before required.

____

____

2. Boiler is used to convert the water into high pressure steam.

____

____

3. It is attached to the cooling tower.

____

____

4. Feed water pump is used to transport the feed water to the boiler.

____

____

5. Generator is coupled with the turbine rotor and converts the mechanical energy of the turbine to the electrical energy.

____

____

Part Three

☞ **Text B**

New Technology Helps Researchers Decrease Emissions at Coal Plants

Researchers at Zhejiang University aim to further lower the emissions of coal-fired power plants by improving the ultralow emission technology they have developed. Their system can effectively reduce the emissions of pollutants such as sulfur dioxide, sulfur trioxide and heavy metals such as mercury.

Coal is the main source of energy in China. According to the International Energy Agency, the country has been the world's largest producer of coal since 1985 and produced nearly half the world's coal in 2014. Pollutants such as nitrogen oxides, particulate matter and sulfur dioxide emitted when coal is burnt are major causes of atmospheric pollution, including smog.

The team at Zhejiang University has been working with Zhejiang Energy Group to apply the technology. In January 2018, the project earned the first prize in the National Technology Invention Award. The Ministry of Science and Technology said the system had drastically reduced the pollutants emitted by coal-fired power plants nationwide while boosting their profits by 1.19 billion yuan ($173 million) from 2014 to 2017.

(168 words)

参考译文

新技术帮助研究人员减少煤电厂污染物的排放

浙江大学的研究人员致力于通过完善他们研发的极低排放技术，从而进一步降低燃煤电厂污染物的排放。这些研究人员研发的系统可以有效减少二氧化硫、三氧化硫以及水银这样的重金属污染物的排放。

煤是中国能源的主要来源。根据国际能源署的数据，从 1985 年以来中国已经成为世界上最大的煤炭生产国，2014 年生产的煤已经接近世界产量的一半。煤燃烧时散发出来的氧化氮、微颗粒物质和二氧化硫是包括雾霾在内的大气污染的主要原因。

浙江大学的研究团队一直在与浙江能源集团合作应用该技术。2018 年 1 月，这个研究项目获得国家技术发明奖一等奖。科技部表示这个系统已经极大地减少了全国燃煤电厂排放的污染物，并且使 2014 到 2017 年间电厂利润增加了十一亿九千万元（折合一亿七千三百万美元）。

Words & Expressions	
ultralow [ˌʌltrə'ləʊ]	*adj.* 极低的
emission [ɪ'mɪʃən]	*n.*（光、热等的）发射，散发；排放；发行
pollutant [pə'luːtnt]	*n.* 污染物
burn [bɜːn]	*vt.* 燃烧；烧毁，灼伤；激起……的愤怒 *vi.* 燃烧；烧毁；发热
smog [smɒg]	*n.* 雾霾
emit [ɪ'mɪt]	*vt.* 发出，放射；发行；发表
boost [buːst]	*vt.* 促进；增加；支援 *vi.* 宣扬
atmospheric pollution	大气污染
heavy metal	重金属
particulate matter	微颗粒物质
the National Technology Invention Award	国家技术发明奖
The Ministry of Science and Technology	科技部

Notes

1. aim to: 旨在；致力于

   e.g. What he has done aims to have a happy life.

   他所做的都是为了过上幸福的生活。

2. the International Energy Agency: 国际能源署，简称 IEA，是总部设在法国巴黎的政府间组织，由经济合作发展组织为应对能源危机于 1974 年 11 月设立。国际能源署致力于预防石油供给的异动，同时亦提供国际石油市场及其能源领域的统计情报。
3. Zhejiang Energy Group: 浙江能源集团，成立于 2001 年 2 月，是经浙江省人民政府批准、以原浙江省电力开发公司和浙江省煤炭集团公司为基础组建而成的省级能源类国有大型企业，主要从事能源建设、电力热力生产、煤矿投资开发、煤炭流通经营、天然气开发利用、能源服务和金融地产等业务。
4. the National Technology Invention Award: 国家技术发明奖，授予运用科学技术知识做出产品、工艺、材料及其系统等重大技术发明的中国公民。

Exercises

Task 1 Fill in the blanks with the right words about atmospheric pollution in Text B.

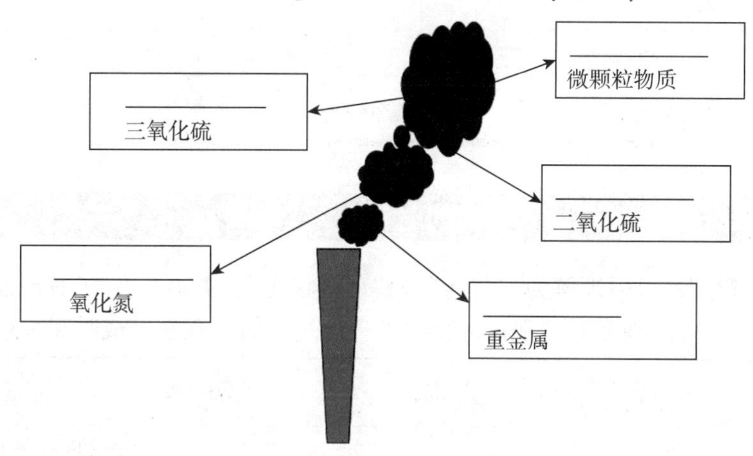

Task 2 Match the English words with the right Chinese.

1. emit	促进；增加
2. pollutant	发出，放射
3. boost	雾霾
4. ultralow	污染物
5. smog	极低的

Task 3 Translate the following sentences into Chinese.

1. Researchers at Zhejiang University aim to further lower the emissions of coal-fired power plants.

____

____

2. Their system can effectively reduce the emissions of pollutants such as sulfur dioxide, sulfur trioxide and heavy metals such as mercury.

____

____

3. Coal is the main source of energy in China.

____

____

4. Pollutants such as nitrogen oxides, particulate matter and sulfur dioxide emitted when coal is burnt are major causes of atmospheric pollution, including smog.

____

____

5. In January 2018, the project earned the first prize in the National Technology Invention Award.

____

____

Section C Advanced Training

Part One Grammar: 动词的时态

英语的“时态”，可以理解为“时间”+“动词状态”。英语共有16种时态形式，最常用的时态有五个：一般现在时、现在进行时、一般过去时、现在完成时和一般将来时。

时间	一般时态	进行时态	完成时态	完成进行时态
现在	一般现在时 watch watches	现在进行时 is/am/are watching	现在完成时 has/have watched	现在完成进行时 has/have been watching
过去	一般过去时 watched	过去进行时 was/were watching	过去完成时 had watched	过去完成进行时 had been watching
将来	一般将来时 shall/will watch	将来进行时 shall/will be watching	将来完成时 shall/will have watched	将来完成进行时 shall/will have been watching
过去将来	一般过去将来时 should/would watch	过去将来进行时 should/would be writing	过去将来完成时 should/would have watched	过去将来完成进行时 should/would have been watching

e.g. My uncle is working at a thermal power plant in Nanning now.
我叔叔现在南宁一家火电厂工作。
My uncle worked at a thermal power plant in Nanning last year.
我叔叔去年在南宁一家火电厂工作。
My uncle has worked at a thermal power plant in Nanning for 3 years.
我叔叔在南宁一家火电厂已经工作了3年。
My uncle will work at a thermal power plant next month.
我叔叔下个月将在南宁一家火电厂工作。
My uncle had finished his work at the thermal power plant before he went home yesterday.
昨天我叔叔在回家前把电厂的工作都完成了。
When my uncle went to the office, his colleague was writing a plan yesterday.
昨天当我叔叔到办公室的时候，他的同事正在写方案。

Exercises

Choose the appropriate answer from the three choices marked A, B and C.

1. My class ____________ visit the Mashi Hydropower Station.
   A. will go to　　B. goes　　C. go
2. My mother ____________ me last night.
   A. call　　B. called　　C. is calling
3. My father ____________ in Laibin Thermal Power Station A for 25 years.
   A. works　　B. has worked　　C. would work
4. When I come home, my mother ____________ .
   A. is cooking　　B. cooks　　C. cooked
5. Please give it to her as soon as she ____________ home.
   A. will come　　B. came　　C. comes
6. This thermal power station ____________ in March 1985.
   A. will establish　　B. establishes　　C. was established

Part Two Writing

Invitation Letter 邀请函

邀请函用于重要会议、开幕典礼、正式晚宴。在写邀请函的时候，格式是非常重要的，邀请函一般由标题、称谓、正文、落款组成。

Sample

Invitation Letter

Jun 27, 2019

Dear Mr. Brown,

Thank you for your letter of Jun 25, 2019. I'm glad that you are also going to Beijing next month. It would be a great pleasure to meet you at the exhibition trade fair.

Our company is having a reception at Beijing Hotel on the evening of July 5, 2019 and I would be very pleased if you could attend.

I look forward to hearing from you soon.

Yours sincerely,
Lucy Hanks

☞**Task**

假设你是来宾电厂B厂的职工，电厂将于6月21日在厂内举行建厂20周年纪念活动，请你写一份邀请函邀请嘉宾参加纪念活动的晚宴。

Section D An English Song

☞ **Task** Listen to the song and fill in the blanks with the missing words you have just heard, and then sing along with it.

Li Bai

陈嬛

Some people walk about this life looking for something special
大部分人忙碌地寻找一种特别的人生
I've turned around and realized I'm losing myself
我辗转反侧才发现我已开始迷失自己
Some movies I have watched but never understood
我看了几部电影却始终没懂
① ________________

Get busy dressing up, putting on your best masks ,looking smart
依旧是终日忙碌着戴上面具，打扮自己，让自己看起来很牛
Sometimes I'd try some tequila, whisky and vodka
有时候我会试试用龙舌兰、威士忌、和伏特加来麻醉自己
Imaging living some kind of beauty in a lie
想象着在一种谎言中活出一点精彩
It's like everyone you met you call him "Darling"
你一天一口一个"亲"
A fashion never out of trend
仿佛那是永不褪色的流行词汇
Perhaps I'd rather take up a skill or two
我看我还是去学两门真功夫算了
Nothing goes oh
别的都没用
Give me one more try, I wanna be Li Bai
要是能重来，我要做李白
The things I did a millennium ago, they would never die
几百年前做的好坏，永不会逝去

Give me one more try，I wanna be Li Bai
要是能重来，我要做李白
The poems I would write they are so fine, everyone would like
写出的诗那么漂亮，谁都会喜爱
② ______________________________

Finding another way to get by，living out a brighter life
找寻另一种存在的方式，活出更多精彩
③ ______________________________

Sometimes I'd try some tequila, whisky and vodka
有时候我会试试用龙舌兰，威士忌，和伏特加来麻醉自己
Imaging living some kind of beauty in a lie
想象着在一种谎言中活出一点精彩
It's like everyone you met you call him 'Darling'
你一天一口一个"亲"
④ ______________________________

Perhaps I'd rather take up a skill or two
我看我还是去学两门真功夫算了
Nothing goes
别的都没用

Background Tip:

陈嬛是一位从华尔街走入马戏团再回归中国艺术界的追梦沙画家和唱作人。这首歌是翻唱，不过歌词仍然是陈嬛独立创作的。《李白》的原唱是李荣浩。

Unit 5

Hydropower Plant

水电厂

Section A Professional Background

Part One Lead-in

☞ **Task** Study the pictures and discuss the questions below.

Questions:

1. Do you know the first hydropower station in Guangxi? Where is it?
2. Which is the largest hydropower station in China?
3. What equipment do you know in the last three pictures?

Cues:
water, hydropower station, reservoir, headwater, head pond, dam, intake, powerhouse, forebay, hydroturbine, hydro generator

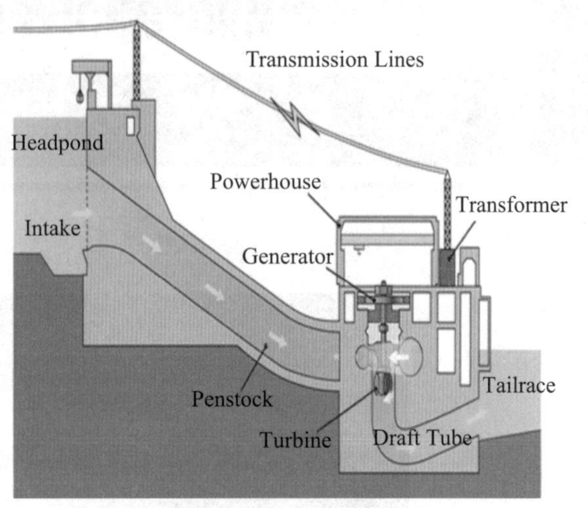

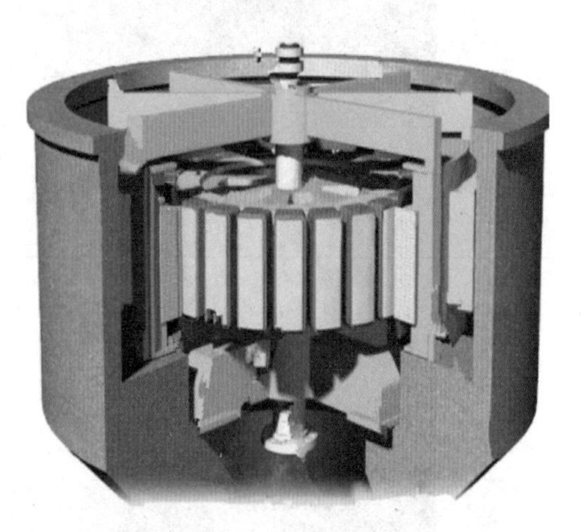

Part Two Listening & Speaking

Listen to the following dialogues and fill in the blanks.

Dialogue I

M: Jenny, do you ___________ the first hydropower station in Guangxi?

W: Yes, Guangxi Yantan Hydropower Station is the first one.

M: When was it built?

W: It was built in March, 1985.

M: When was it ___________?

W: It was finished successfully ___________ of the river (截流) in November, 1987.

M: When could it generate electricity?

W: The first generating ___________ went into generating electricity in 1992.

M: How many ___________ are there in the power plant?

W: There are about 435 ones.

☞Dialogue 2

W: Tom, do you know the first hydropower plant in the ___________?

M: Yes, in 1878, France built the world's first hydropower plant.

W: Which is the ___________ one in China?

M: The ___________ Gorges Dam (三峡水电站).

W: When was it built?

M: In 1994.

W: What about the ___________ capacity (装机容量)?

M: About 17.86 million kW.

W: I am looking forward to ___________ it.

M: So am I.

Section B Text Learning

Part One Lead-in

1. Which is the first hydropower station in China?
2. How many types of hydropower plants are there?
3. Are there any disadvantages of hydropower generation?

Part Two

Text A

Hydroelectric Power Plant (Hydropower Station)

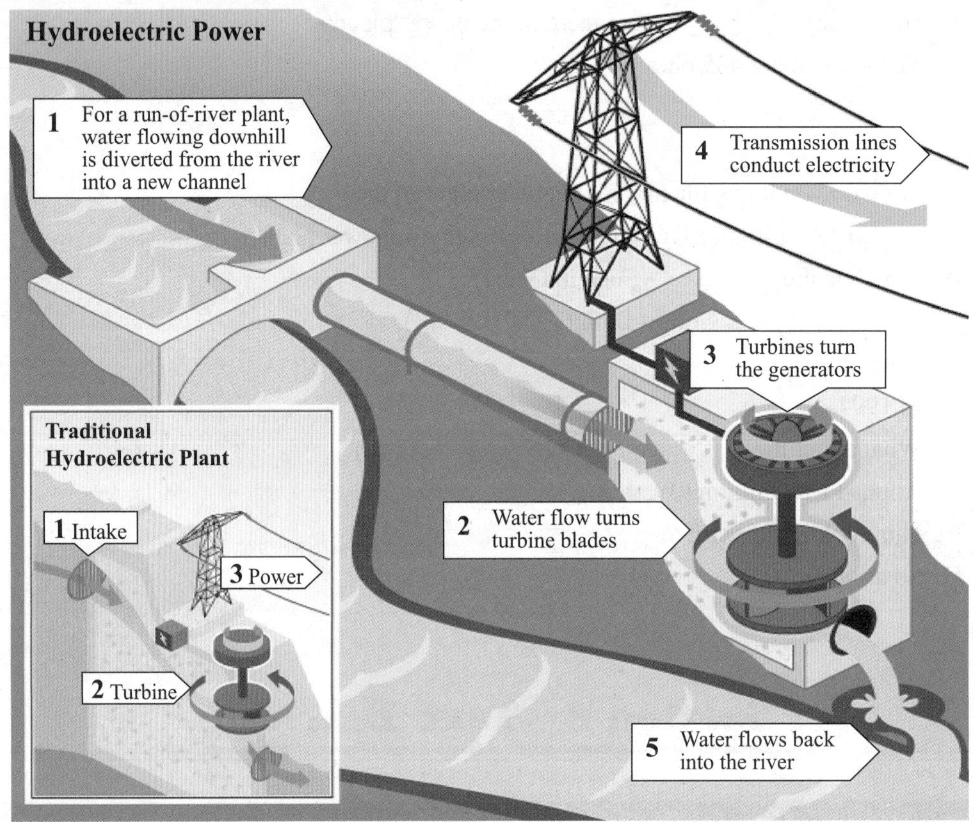

Fig. 5.1

Hydropower generated about 19.4% of the nation's energy in 2015. Water flowing in the river is comprised of kinetic energy and potential energy that is utilized to produce electricity. The important components of the hydroelectric power plant can be seen in Fig. 5.2.

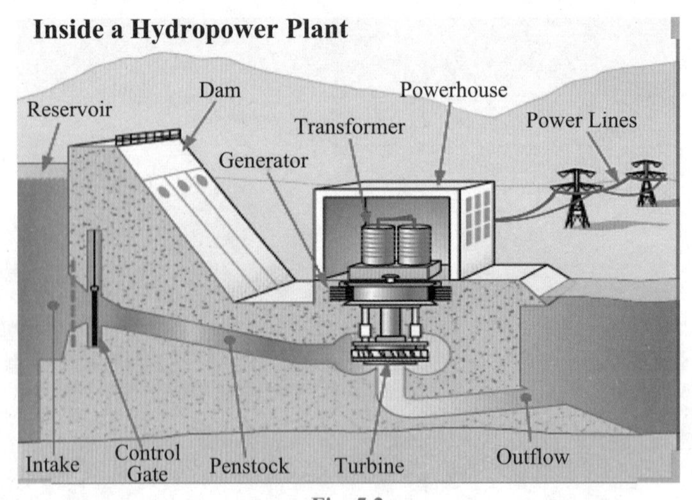

Fig. 5.2

Water reservoir (Fig. 5.3) is the place behind the dam where water is stored. The dam is built on a large river to increase the height of the water level (increase in the potential energy) to increase the reservoir capacity.

Intake or Control Gates (Fig. 5.3) were built on the inside of the dam. The water from reservoir is released and controlled through these gates. When the control gates are opened, the water flows due to gravity through the penstock and towards the turbines. The water flowing through the gates possesses potential energy as well as kinetic energy. (Fig. 5.3)

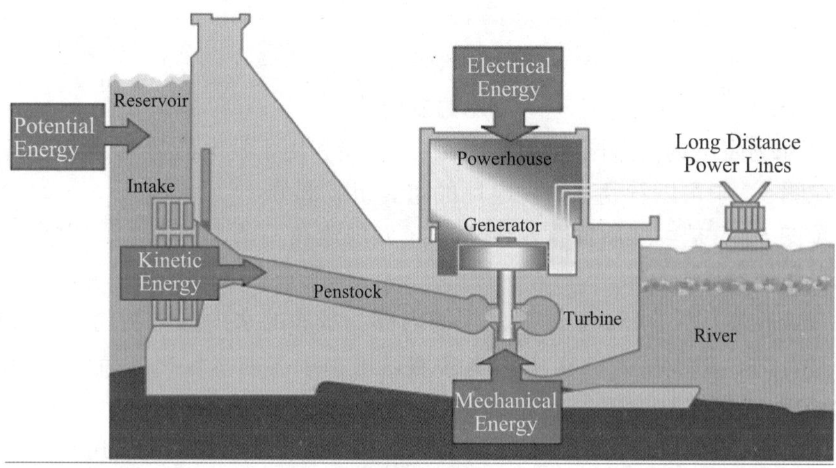

Fig. 5.3

Penstock (Fig. 5.3) is the long pipe or the shaft that carries the water flowing from the reservoir towards the power generation unit.

Water turbine (Fig. 5.3) is to convert the kinetic energy of the water into the mechanical energy to produce the electric power. Water flowing from the penstock enters the power generation unit and exerts the force of the water on turbine blades to make them rotate.

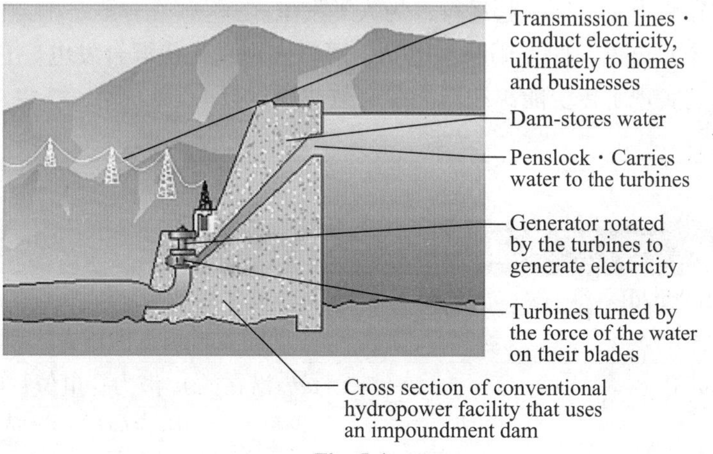

Fig. 5.4

Generator is mechanically coupled to the turbine shaft. When the turbine blades are rotated, the blades drive the generator to generate electricity. A transformer is used to step up the electricity from the generator before transmission.

A power house consists of two main parts, a sub-structure to support the hydraulic and electrical equipment and a superstructure to store and protect this equipment. All the operating equipment is in the superstructure of the powerhouse. These important components work together in a hydropower plant to produce electricity for the human beings.

In China there is a hydropower plant, named Longyangxia hydropower plant, which can achieve water-pv hybrid generation. It avoids the limitations of hydropower generation. It realizes the effective coordinated operation of traditional energy and new energy.

(337 words)

参考译文

水力发电厂

2015 年水力发电约占全国能源的 19.4%。动能和用于发电的势能是由河流中的水流组成的。在图 5.2 中可以看到水力发电厂的重要组成部分。

水库（图 5.3）是大坝后面蓄水的地方。大坝建在一条大河流上，以增加水位的高度和增加水库的容量。

进水口或控制门（图 5.3）建在大坝的内部。水库的水由这些闸门释放和控制。当控制闸门打开时，由于重力的作用，水会通过进水渠流向水轮机。流经阀门的水流具有势能和动能。（图 5.3）

进水渠（图 5.3）是一条长管道或输水管，它承载着从水库流向发电机组的水。

水轮机（图 5.3）是将水的动能转换成机械能来产生电能。水流从进水渠进入发电机组，并利用水力推动水轮机叶片旋转。

发电机与水轮机耦合。当水轮机叶片旋转时，它驱动发电机发电。发电机发出的电经由变压器进行变压再传输。

发电站的动力室主要由两个部分组成，一个是支撑水力和电气设备的子结构，一个是用来储存和保护这些设备的上层建筑。所有的设备操作都在发电站的上层建筑中。这些重要的组成在水力发电厂一起运行，为人类发电。

中国有一个水电站，叫龙阳峡水电站，可以实现水光伏混合发电。它避免了水力发电的局限性。它实现了传统能源和新能源有效协调运行。

Words & Expressions

comprised [kəmp'raɪzd]	*adj.* 包含的 （动词 comprise 的过去式和过去分词）
potential [pə'tenʃl]	*adj.* 潜在的 *n.* 潜力；电位；电势
utilize ['juːtəlaɪz]	*vt.* 〈美〉利用或使用 =〈英〉utilise
reservoir ['rezəvwɑː(r)]	*n.* 水库；储藏；蓄水池
dam [dæm]	*n.* 水坝；堤 *v.* 筑坝
located [ləʊ'keɪtɪd]	*adj.* 处于；位于

height [haɪt]	*n.* 高度；高处
capacity [kə'pæsəti]	*n.* 容量；容积；能力
intake ['ɪnteɪk]	*n.* 入口；通风口；吸入；招收
flow [fləʊ]	*n.* 流动；流；流量 *vi.* 流动；涌出；飘动 *vt.* 淹没
gravity ['grævəti]	*n.* 重力；庄重
penstock ['penstɒk]	*n.* 水门；水道；水闸
possess [pə'zes]	*vt.* 拥有；持有；支配
shaft [ʃɑ:ft]	*n.* 轴；柄；矛
coupled ['kʌpld]	*adj.* 成对的（连接的；共轭的；联系的）
powerhouse ['paʊəhaʊs]	*n.* 发电站；动力室
substructure ['sʌbstrʌktʃə]	子结构
superstructure ['su:pəstrʌktʃə(r)]	*n.* 〈政〉上层建筑；〈建〉上层结构
hydraulic [haɪ'drɔ:lɪk]	*adj.* 水力的；水压的；液压的

Notes

1. reservoir capacity: 贮水量，水库容量；reservoir water capacity: 水库蓄水量
2. control gate: 控制闸门
3. power-generation unit: 发电机组
   e.g. Now, hydropower power generation unit has developed into MW level from kW level.
   目前，水力发电机组已由千瓦级发展到兆瓦级。

Exercises

Task 1 Answer the following questions according to the text.

1. What percentage of the nation's energy did hydropower generate in 2015?
   ____
2. What is penstock?
   ____
3. Why is the water in the reservoir located higher than the rest of the dam structure?
   ____
4. What's the function of water turbine in a hydropower plant?
   ____
5. How many main parts does a power house consist of ?
   ____

Task 2 Change the word formation.

Before	After	Meaning
comprised *adj.*		*v.* 构成；包含
located *adj.*		*v.* 找出；使坐落于

flow *v.*		*adj.* 流动的；平滑的
capacity *n.*		*adj.* 有能力的；熟练的；胜任的
hydraulic *adj.*		*n.* 水疗院；水利

☞ **Task 3** Complete the following sentences with the proper form of the words from Task 2.

1. Yantan Hydropower Station is __________ in the middle of the Red River of the Yantan Town of Dahua.
2. What is the carrying __________ of this ship?
3. In the __________ plant, the average power generation is 52.26 billion kilowatt hours (千瓦时).
4. The river __________ into the sea.
5. A power house is __________ of the substructure and the superstructure.

☞ **Task 4** Translate the following sentences into Chinese.

1. Water flowing in the river is comprised of kinetic energy and potential energy that is utilized to produce electricity.

____

____

2. Water reservoir is the place behind the dam where water is stored.

____

____

3. The water flowing through the gates possesses potential as well as kinetic energy.

____

____

4. Generator is mechanically coupled to the turbine shaft.

____

____

5. These important components work together in a hydropower plant to produce electricity for the human beings.

____

____

Part Three

☞ **Text B**

Environmental Impacts of Hydroelectric Power

Land Use

Flooding land for a hydroelectric reservoir has an extreme environmental impact: it destroys forest, wildlife habitat, agricultural land, and scenic lands. In many instances, such as the Three

Gorges Dam in China, entire communities also had to be relocated to make way for reservoirs.

Wildlife Impacts

Though there are a variety of methods to minimize the impact, fish and other organisms can be injured and killed by turbine blades. Apart from direct contact, there can also be wildlife impacts both within the dammed reservoirs and downstream from the facility.

Reservoir water is typically low in dissolved oxygen and colder than normal river water. When this water is released, it could have negative impacts on downstream plants and animals. In addition, if too much water is stored behind the reservoir, segments of the river downstream from the reservoir can dry out.

Global Warming Emissions

Global warming emissions are produced during the installation and dismantling of hydroelectric power plants, but recent research suggests that emissions during a facility's operation can also be significant. Such emissions vary greatly depending on the size of the reservoir and the nature of the land that was flooded by the reservoir.

(194 words)

参考译文

水力发电对环境的影响

土地使用

为修建水电站水库而淹没土地会对环境造成极端影响：它会破坏森林、野生动物栖息地、农业用地和风景区。在许多情况下，比如中国的三峡大坝，整个社区也不得不搬迁，为水库让路。

野生动物的影响

虽然有各种各样的方法可以将影响降到最低，但鱼和其他生物可能会被涡轮机叶片伤害和杀死。除了直接接触外，水坝内及设施下游亦可能对野生动物造成影响。

水库的水通常溶解氧含量低，比正常的河水更冷。当这些水被释放出来时，可能会对下游的动植物造成负面影响。此外，如果水库后面储存了太多的水，水库下游的部分河流可能会干涸。

全球变暖排放

全球变暖的排放是在水电站的安装和拆除过程中产生的，但最近的研究表明，设施运行过程中的排放也可能是巨大的。这种排放会因水库的大小和被水库淹没的土地的性质而发生很大的变化。

Words and Expressions

habitat ['hæbɪtæt]	*n.* [生态] 栖息地，产地
agricultural [ˌægrɪ'kʌltʃərəl]	*adj.* 农业的；农艺的
community [kə'mju:nəti]	*n.* 社区；[生态] 群落；共同体；团体

minimize ['mɪnɪˌmaɪz]	*vt.* 使减到最少；小看 *vi.* 最小化
organism ['ɔːgənɪzəm]	*n.* 有机体；生物体；微生物
facility [fə'sɪləti]	*n.* 设施；设备
dissolve [dɪ'zɒlv]	*vt.* 使溶解；使分解 *vi.* 溶解；解散
oxygen ['ɒksɪdʒən]	*n.* [化学] 氧气，氧
negative ['negətɪv]	*adj.* 负的；消极的；否定的；阴性的
emission [ɪ'mɪʃən]	*n.* 排放；排放物；（光、热等的）发射
dismantle [dɪs'mæntl]	*vt.* 拆除；取消；解散
significant [sɪg'nɪfɪkənt]	*adj.* 重大的；有意义的；值得注意的
make way for	让路；为……开路
dry out	变干；戒酒
depend on	取决于；依赖；依靠

Notes

1. Three Gorges Dam: 三峡大坝
2. In many instances: 在许多情况下
   instance 常见的用法是 for instance，意为“例如”，相当于: for example。
3. a variety of: 种种；各种各样的……
   e.g. A variety of investments can lower your risks, so there is no need to put all your eggs into one basket.
   一个投资品种可以降低风险，所以没有必要把所有的鸡蛋放在一个篮子里。

Exercises

Task 1 Write down the words according to their explanations.

1. ____________	*n.* a group of people living in a particular local area
2. ____________	*v.* to mix with a liquid and become part of it
3. ____________	*adj.* harmful, unpleasant, or not wanted
4. ____________	*n.* a gas or other substance that is sent into the air
5. ____________	*adj.* important in effect or meaning

Task 2 Match the following impacts and their corresponding sequences.

1. Land Use
2. Wildlife Impacts
3. Global Warming Emissions

A. Fish might be injured and killed by turbine blades.
B. River downstream from the reservoir can dry out.
C. It destroys forest, wildlife habitat.
D. Facility's operation can produce emission.
E. Installation and dismantling of hydroelectric power plants will produce emissions.

☞ **Task 3** Translate the following sentences into English.

1. 整个社区也不得不搬迁，为水库让路。

____

2. 有各种各样的方法可以将影响降到最低。

____

3. 水库的水通常比正常的河水更冷。

____

4. 水库后面储存了太多的水。

____

5. 设施运行过程中的排放也可能是巨大的。

____

Section C Advanced Training

Part One Grammar: 动名词

动名词是动词的另一种非限定形式，它在句子中起名词的作用。动名词或动名词短语在句子中可以作主语、表语、宾语或者作介词的宾语。

1. 作主语

e.g. Flowing water in the penstock is to the turbine. 管道中的流水流向水轮机。

2. 作宾语

1) 动词后加动名词 doing (*v.* + doing sth.) 作宾语

consider 认为	admit 承认	avoid 避免	appreciate 感激，赞赏
deny 否认	enjoy 喜欢	finish 完成	prevent 阻止
complete 完成	keep 继续	delay 耽误	mind 介意
escape 逃脱	miss 想念	postpone 推迟	imagine 想象

e.g. I don't mind telling you how to operate this control gate.

我不介意告诉你如何操作这个控制闸门。

2) 词组后接 doing 作宾语

be/get used to 习惯于	busy 忙于	prefer…to 宁愿……不愿
look forward to 期望；盼望（to 为介词）	It's worth… 它值得	can't help 忍不住
It's no use /good 做……是没有用的	be fond of 喜爱	keep on 继续

e.g. He is used to drinking tea in the office. 他习惯了在办公室喝茶。

3. 作表语

e.g. Her job is monitoring the operation of hydro turbine. 她的工作是监控水轮机的运行。

Exercises

Choose the appropriate answer from three choices marked A, B and C.

1. ____________ on the dam of the hydropower plant is his job.
   A. Walk B. Walks C. Walking
2. It's worth ____________ the film.
   A. seeing B. to see C. sees
3. She enjoys ____________ her job in the Longyangxia hydropower plant.
   A. did B. do C. doing
4. His ideal is ____________ that hydropower power plant.
   A. entering B. enters C. entered
5. He was afraid ____________ for being late.
   A. of being seen B. of seeing C. to be seen
6. I suggests ____________ the Longtan hydropower plant next week.
   A. visits B. visiting C. to visit

Part Two Writing

备忘录（Memo）是单位内部传递信息的文件，格式与外部通知（Notice）不同，备忘录包括抬头和正文，其中抬头的内容及格式如下：

MEMO
To:
From:
Date:
Subject:
Content（正文）:

Sample

MEMO
To: All department heads
From: Mike Jackson
Date: June 11, 2018
Subject: Discussion on the Sales Plan of the 4th Quarter, 2018
The sales department has made a new plan for the 4th quarter of 2018, a meeting will be held to discuss the plan in Conference Room 1 at 2 : 30 p.m. on June 11, 2018. All department managers are required to be present on time.

☞ **Task**

说明：假如你是麻石电厂厂区主管王明东，请以王明东的名义按照下面的格式和内容给本部门员工写一个内部通知 Memo。

主题：讨论 2018 年度第三季度（the third quarter）的安全工作

通告时间：2018 年 6 月 12 日

内容：本部门已制订 2018 年第三季度的安全工作。将于 2018 年 6 月 18 日下午 4：30 在厂区 1 号会议室开会，布置工作，并希望本部门员工按时来参加。

Section D An English Song

☞ **Task** Listen to the song and fill in the blanks with the missing words you have just heard, and then sing along with it.

See You Again

Wiz Khalifa, Charlie Puth

It's been a long day without you my friend 没有老友你的陪伴 日子真是漫长
And I'll tell you all about it when I see you again 与你重逢之时 我会敞开心扉倾诉所有
We've come a long way from where we began 回头凝望 我们携手走过漫长的旅程
Oh I'll tell you all about it when I see you again 与你重逢之时 我会敞开心扉倾诉所有
When I see you again 与你重逢之时
Damn who knew all the planes we flew 谁会了解我们经历过怎样的旅程
Good things we've been through 谁会了解我们见证过怎样的美好
That I'll be standing right here 我都会在这里
Talking to you about another path 与你聊聊另一种选择的可能
I know we loved to hit the road and laugh 我懂我们都偏爱速度与激情
But something told me that it wouldn't last 但有个声音告诉我这美好并不会永恒
Had to switch up look at things 我们得变更视野
① ____________________

Those were the days hard work forever pays 有付出的日子终有收获的时节
Now I see you in a better place 此刻 我看到你走进更加美好的未来
How could we not talk about family 当家人已是我们唯一的牵绊
② ____________________

Everything I went through 无论历经怎样的艰难坎坷
you were standing there by my side 总有你相伴陪我度过
And now you gonna be with me for the last ride 最后一段征程 我更需要你的相伴
It's been a long day without you my friend 没有老友你的陪伴 日子真是漫长
And I'll tell you all about it when I see you again 与你重逢之时 我会敞开心扉倾诉所有
We've come a long way from where we began 回头凝望 我们携手走过漫长的旅程
Oh I'll tell you all about it when I see you again 与你重逢之时 我会敞开心扉倾诉所有
When I see you again 与你重逢之时
First you both go out your way 一开始你们总是追随你们心中的步伐
And the vibe is feeling strong and what's 热忱累积 信念不变
Small turn to a friendship a friendship 渺小的世界见证这段深情厚谊
Turn into a bond and that bond will never 深厚的友情变成血浓于水的感情
Be broke and the love will never get lost 此情不变 此爱难逝
And when brotherhood come first then the line 莫逆之交的我们绝不会背叛彼此
Will never be crossed established it on our own 只因这深情厚谊基于我们真实意愿
When that line had to be drawn and that line is what 这友谊让我们肝胆相照 荣辱与共
We reach so remember me when I'm gone 即便我离去 也请将我铭记
How could we not talk about family 当家人已是我们唯一的牵绊
When family's all that we got? 我们怎么能忘却最可贵的真情
③ ______________________________

You were standing there by my side
总有你相伴陪我度过
And now you gonna be with me for the last ride
最后一段征程 我更需要你的相伴
Let the light guide your way hold every memory
就让那光芒引导你的前路 铭记与我的曾经
As you go and every road you
无论你选哪一条路
Take will always lead you home
那都会通向你的家
Hoo~ 吼 ~~
It's been a long day without you my friend.
没有老友你的陪伴日子真是漫长。
And I'll tell you all about it when I see you again
与你重逢之时 我会敞开心扉倾诉所有

We’ve come a long way from where we began
回头凝望 我们携手走过漫长的旅程
Oh I’ll tell you all about it when I see you again
与你重逢之时 我会敞开心扉倾诉所有
When I see you again
与你重逢之时
Again
重逢之时
When I see you again see you again
与你重逢之时 重逢之时
④ ____________________

Background Tip:

See You Again 是由美国著名说唱歌手 Wiz Khalifa 和新人 Charlie Puth 共同演唱的一首歌曲。这首歌收录于电影《速度与激情 7》的原声带中，也是电影的片尾曲兼主题曲。2015 年该歌曲获得青少年选择奖最佳 R&B 嘻哈歌曲、最佳影视歌曲奖。

Unit 6

Concentrated Solar Power

聚光太阳能发电

Section A Professional Background

Part One Lead-in

☞ **Task** Study the pictures and discuss the questions below.

Questions:

1. What can you see in these pictures?
2. Are there any concentrated solar power plants in Guangxi? Please talk about them.
3. How many kinds of solar power generation are there in the world? What are they?

Cues:
plant, sunlight, solar panel, lamp, mirror, heliostat, solar tower, solar panel

Part Two Listening & Speaking

Listen to the following dialogues and fill in the blanks.

☞Dialogue I

W: Where did you go during this summer vacation, Henry?

M: I went to the countryside during the summer vacation.

W: What did you think of it?

M: I really enjoyed the __________ weather and the __________ scenery.

W: Really? But I heard that the countryside was very __________.

M: No, the village has been carrying out __________ activities for several years, so you can't find any garbage on the road. I suggest you should participate in the summer countryside activities.

W: OK, I will __________ them next summer vocation.

M: You'll find it worthy.

☞Dialogue 2

W: Hey Rob, what did you __________ during this summer holiday?

M: I participated in the summer vocation countryside activities. It could __________ myself and prepare for the future job.

W: Did you find anything __________ ?

M: Yes, I found there were some solar street lamps on the road.

W: As I know, __________ percent of China's energy relies on coal. Coal and oil account for more than 90%.

M: You know the use of coal and oil will pollute the environment and they __________ to non-renewable energy sources, so green and low-carbon industries such as new energy equipment (新能源装备), and new energy vehicles (新能源汽车) have been advocated.

W: Solar energy is one of inexhaustible energy resources.

M: Yes, everyone likes the green life.

Section B　Text Learning

Part One　Lead-in

1. Which is the largest solar power station in China?
2. Where is the world's largest solar power station located in?
3. Can new energy be applied to our daily life at present?

Part Two

Text A

Concentrated Solar Power

Solar energy is one of inexhaustible energy resources. According to National Energy Administration, China had added 34.54GW of photovoltaic power generation capacity, with a total installed capacity of 77.42GW by the end of 2016. Currently, solar power accounts for just 1 percent of China's total power generation in 2016. The National Energy Administration will have planned to add 110 GW of photovoltaic power generation capacity by 2020. By 2030, China's share of non-fossil fuel in total energy consumption will have risen from 11 percent to 20 percent.

Solar power converts sunlight into electricity either directly using photovoltaics (PV), or indirectly using concentrated solar power. Concentrated solar power systems use lenses or mirrors and tracking systems to focus a large area of sunlight into a small beam. Photovoltaic cells convert light into an electric current by the photovoltaic effect.

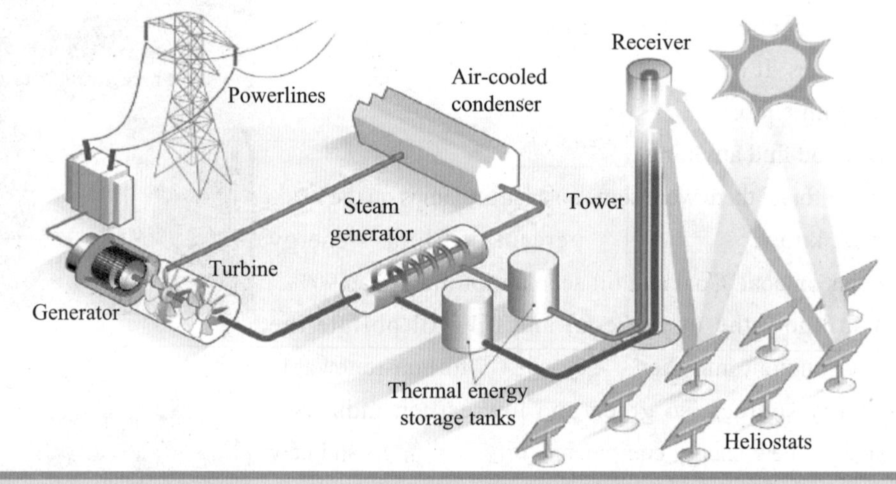

In a CSP system with thermal energy storage, the heat transfer medium, such as molten salt, retains heat so well that it enables the plant to generate electricity for hours when the sun is not shining.

Concentrated solar power (CSP) plants use mirrors to concentrate the energy from the sun to drive traditional steam turbines or engines to create electricity. The thermal energy concentrated in a CSP plant can be stored and used to produce electricity when it is needed in the daytime or evening.

A concentrated solar power (CSP) plant contains heliostats, central receiver, power tower, dish-engine, turbine, and generator and so on.

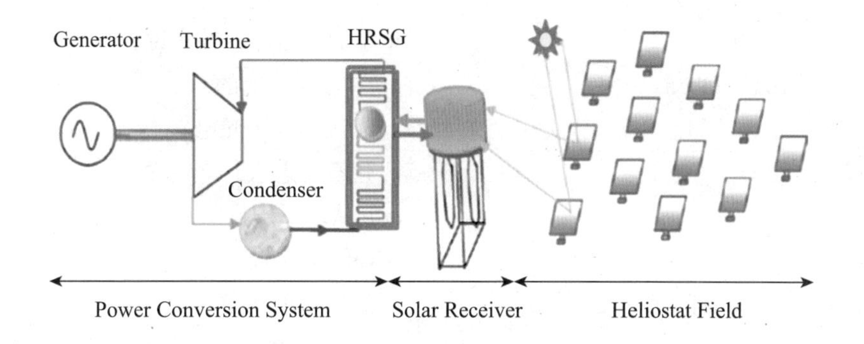

In Dunhuang, there is a Molten Salt Tower Photothermal Power Station, which is a kind of concentrated solar power (CSP) station. This station is the third power station in the world and the first in Asia that can realize 24-hour continuous generation of molten salt tower type photothermal power. This power station adopts 100% solar energy and does not need other energy to refuel.

(275 words)

参考译文

聚光太阳能发电

太阳能是取之不尽用之不竭的能源之一。据国家能源局网站称，截至 2016 年年底，中国光伏发电新增装机容量 34.54GW，累计装机容量 77.42GW。当前，太阳能仅占中国全年总发电量的 1%。国家能源局计划至 2020 年增加光伏发电新增装机容量 11 000 万千瓦；到 2030 年，中国非化石能源燃料占能源消费总量的比重将从 11% 增加至 20%。

太阳能就是把阳光转化为电能，或直接使用光伏（PV），或间接使用集中太阳能。集中式太阳能发电系统使用透镜或镜子和跟踪系统将大面积的阳光聚焦成一束小光束。光电电池利用光电效应将光转换成电流。

聚光式太阳能发电厂（CSP）利用镜子将来自太阳的能量集中起来，驱动传统的汽轮机或发动机发电。集中在 CSP 电厂的热能可以储存起来，在需要的时候，白天黑夜都可以用来发电。

聚光式太阳能发电（CSP）装置包括定日镜、中央接收器、发电塔、碟形发动机、涡轮机、发电机等。

敦煌熔盐塔式光热电站，是一座聚光式太阳能发电站。这座电站是全球第三座、亚洲第一座可实现 24 小时连续发电的熔盐塔式光热电站。该电站 100% 采用太阳能，无须其他能源补燃。

Words & Expressions

solar ['səʊlə(r)]	*adj.* 太阳的；太阳能的 *n.* 日光浴室
inexhaustible [ˌɪnɪg'zɔːstəb(ə)l]	*adj.* 用之不竭的；无穷无尽的

add [æd]	*v.* 添加；增加；补充说；继续说
photovoltaic [ˌfəʊtəʊvɒl'teɪɪk]	*adj.* [物] 光电的 *n.* 光伏；太阳光电
fossil ['fɒs(ə)l]	*n.* 化石；老人；老古董 *adj.* 从地下发掘出来的；化石的 fossils （复数）
consumption [kən'sʌmpʃ(ə)n]	*n.* 消费；消耗量
lens [lenz]	*n.* 透镜；镜片
mirror['mɪrə(r)]	*n.* 镜；反射镜；反映；借鉴 *v.* 反映；反射
beam [biːm]	*n.* 梁；光线；平衡木；（电波的）波束 *v.* 照射；发光
cell [sel]	*n.* 元件；[生] 细胞；[电] 电池
heliostat ['hiːlɪə(ʊ)stæt]	*n.* [天] 定日镜；日光反射装置
focus ['fəʊkəs]	*n.* 关注；调焦；中心点（指人或事物） *v.* 集中（注意力、精力等于）；（使）调节焦距；集中（光束于）
central receiver	中央接收器
dish-engine	碟式发电机
non-fossil fuel energy	非化石能源
photovoltaics (PV)	太阳能光伏发电
solar energy	太阳能
The National Energy Administration	国家能源局
tracking system	跟踪系统

Notes

1. account for: 占……比例
2. GW: (gigawatt) 千兆瓦

   e.g. In 2012, Germany had a larger demand for solar photovoltaic power, adding 3 gigawatts (GW).

   在 2012 年，德国太阳能光伏发电需求量增大，增加了 3 000 兆瓦 (GW)。

Exercises

Task I Answer the following questions according to the text.

1. What energy is mentioned in the text?

   ____

2. How many percents of China's total power generation does solar power account for currently?

   ____

3. How many percents will China's share of non-fossil fuel energy consumption have risen from 11% to by 2030?

____

4. How does a concentrated solar power (CSP) plant work?

____

5. What does a concentrated solar power (CSP) plant contain?

____

☞ **Task 2** Change the word formation.

Before	After	Meaning
inexhaustible *adj.*		*n.* 可空竭；可用尽
add *v.*		*adj.* 附加的；更多的
installed *adj.*		*v.* 1. 安装（机器）；设置 2. 建立，设立；修建装置
consumption *n.*		*v.* 消耗；吃；毁灭；烧毁
tracking *n.*		*v.* 跟踪；追踪

☞ **Task3** Complete the following sentences with the proper form of the words from Task 2.

1. Solar energy is __________ energy source.
2. The plant will have planned to __________ 110 GW of photovoltaic power generation capacity by 2020.
3. The Noor-1（一期工程）solar power station will have had an __________ capacity of up to 160 megawatts by 2020.
4. By 2020, non-fossil energy __________ will have accounted for about 15% of total primary energy.
5. Computer-controlled flat mirrors (called heliostats) __________ the sun.

☞ **Task 4** Translate the following sentences into Chinese.

1. By the end of 2016, China had added 34.54GW of photovoltaic power generation capacity.

____

____

2. Concentrated solar power systems use lenses or mirrors and tracking systems to focus a large area of sunlight into a small beam.

____

____

3. Photovoltaic cells convert light into an electric current by the photovoltaic effect.

____

____

4. Concentrated solar power (CSP) plants use mirrors to concentrate the energy from the sun to drive traditional steam turbines or engines to create electricity.

____

____

5. Currently, solar power accounts for just 1 percent of China's total power generation.

____

____

Part Three

Text B

Concentrated Solar Power Projects Fast Gaining Ground

Concentrated solar power generation is making tangible progress in China, with the sector set to be an integral part of the country's current power system soon, said experts. Concentrated solar power, or CSP, is a technology that uses mirrors to concentrate and reflect sunlight to drive traditional steam turbines to create energy.

According to industry newspaper *China Electric Power News*, six CSP projects are likely to be put in operation in 2019. The projects are among the first batch of 20 CSP demonstration plants approved by industry regulator, the National Energy Administration, with planned installed capacity totaling 1.35 gigawatts. Preferential policies from the government helped ensure financial support to these projects.

"CSP plants have great environmental benefits as a type of clean energy compared with fossil fuels. CSP plants also generate more stable power than photovoltaic plants," said Han Xiaoping, chief researcher at energy analysis website China5e. "Moreover, CSP allows power generation at night as current CSP plants are able to store heat from the sun, and convert it into electricity at night, when people come home after a day's work, usually a peak period for electricity," Han said.

Lin Boqiang, head of the China Institute for Studies in Energy Policy at Xiamen University, was more conservative about the industry. "The problem is that the cost of CSP projects is too high, speaking of the land they occupy and the expense of their equipment. That is a problem that hinders the industry's development and a problem many companies are trying to tackle," Lin said.

(254 words)

参考译文

聚光太阳能热发电项目取得飞速发展

专家表示，随着创新技术机组很快将成为中国当前发电系统的一个不可分割的部分，中国的聚光太阳能发电飞速发展，进步明显。聚光太阳能热发电，简称 CSP，是一项利用镜子聚集并反射太阳光以便驱动传统汽轮机产生电能的技术。

根据电力行业报纸《中国电力新闻》，6 个聚光太阳能发电项目极有可能在 2019 年投入使用。这些项目就应用在由电力行业管理部门国家能源局批准的首批 20 个聚光太阳能示范电厂，计划装机容量总数为 1.35 千兆瓦。政府优惠政策给这些项目提供资金支持。

“与石油相比，聚光太阳能发电厂作为一种清洁能源对环境益处巨大。聚光太阳能电厂也能产生比光电池电厂更稳定的电能。”中国能源网 China5e 的首席研究员韩晓平说。“而且聚光太阳能技术可以实现晚上发电，因为目前聚光太阳能电厂可以白天存储太阳的热量，晚上把太阳能量转化为电能。当人们结束一天的工作回到家里，晚上通常会迎来用电的高峰期。”韩晓平表示。

厦门大学中国能源政策研究院林伯强对电力行业则持更保守的态度。林伯强认为目前的问题是聚光太阳能发电项目在电厂的占地和设备的购置费用方面成本太高。这个问题不仅阻碍了电力行业的发展，也是许多公司正在试图解决的问题。

Words & Expressions

Words & Expressions	
concentrate ['kɒnsəntreɪt]	*vi.* 集中；浓缩；全神贯注；聚集 *vt.* 集中；浓缩
tangible ['tændʒəbl]	*adj.* 有形的；切实的；可触摸的
integral ['ɪntɪgrəl]	*adj.* [数学] 积分的；完整的，整体的
reflect [rɪ'flekt]	*vt.* 反映；反射，照出；反省 *vi.* 反射，映现；深思
batch [bætʃ]	*n.* 一批；一炉；一次所制之量
demonstration [ˌdemən'streɪʃən]	*n.* 示范；证明；示威游行
occupy ['ɒkjʊpai]	*vt.* 占据，占领；居住；使忙碌
hinder ['hɪndə]	*vi.* 成为阻碍 *vt.* 阻碍；打扰
tackle ['tækl]	*vt.* 处理；抓住；固定；与……交涉 *vi.* 擒抱摔倒；拦截抢球
chief researcher	首席研究员
preferential policy	优惠政策
fossil fuels	石油

Notes

1. are likely to: 可能

   e.g. CSP is likely to become the major way of power generation in the near future.

   不久的将来聚光太阳能发电可能会成为主要的发电方式。

2. put in operation: 投入使用

   e.g. More CSP plants will be put in operation in China.

   中国将有更多的聚光太阳能发电厂投入使用。

3. a peak period for electricity: 用电高峰期

   e.g. The weekend is usually a peak period of electricity.

   周末通常是用电高峰期。

4. be more conservative about: 对……持更为保守的态度

   e.g. People in local areas are more conservative about applying nuclear for electricity generation. 当地人对使用核能发电态度保守。

5. speak of: 谈到；涉及

   e.g. Speaking of the development of electricity generation, Chinese people have a positive attitude. 对于电力发展，中国人持积极态度。

6. China5e: 中国能源网，提供能源经济、煤炭、电力、油气、新兴能源、节能环保、页岩气、分布式能源等能源行业的能源资讯、能源分析、能源数据、能源研究等能源信息与咨询服务。

Exercises

☞ **Task I** Fill in the blanks with the right words according to the picture.

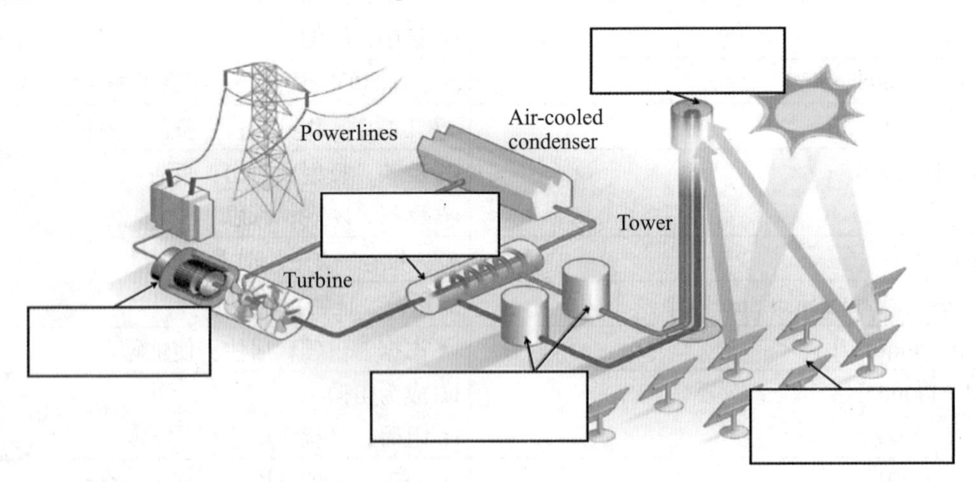

☞ **Task 2** Match the English with the right Chinese words.

1. occupy	反映；反射
2. tangible	占据，占领
3. reflect	集中；浓缩
4. hinder	有形的；切实的
5. concentrate	阻碍；打扰

☞ **Task 3** Translate the following sentences into Chinese.

1. Concentrated solar power, or CSP, is a technology that uses mirrors to concentrate and reflect sunlight to drive traditional steam turbines to create energy.

____

____

2. According to industry newspaper *China Electric Power News*, six CSP projects are likely to be put in operation in 2019.

____

____

3. Preferential policies from the government helped ensure financial support to these projects.

____

____

4. CSP plants have great environmental benefits as a type of clean energy compared with fossil fuels.

____

____

5. The problem is that the cost of CSP projects is too high, speaking of the land they occupy and the expense of their equipment.

____

____

Section C Advanced Training

● Part One Grammar: 动词分词

1. 分词是动词的另一种非限定形式，起到形容词和副词的作用，可以作定语、表语、状语和宾语补足语。分词有现在分词和过去分词两种。现在分词的形式同动名词一样，一般在动词后面加 ing，而过去分词的形式则在一般动词后面加 ed。

2. 分词的作用

1) 分词作定语：

e.g. This is an interesting story. 这是一个有趣的故事。（做前置定语）

The girl reading the newspaper is my sister. 正在看报纸的那个女孩是我妹妹。（做后置定语）

注：被修饰的是 something, anything, everything, nothing 等不定代词时，分词放在被修饰的不定代词后面。

e.g. I have nothing interesting to tell you. 我没有什么有趣的事要告诉你。

拓展：分词和动名词都可以作定语，判断是分词还是动名词，可以根据它们和被修饰词有无逻辑上的主谓关系来判断，有主谓关系的是分词，否则判断为动名词。

a swimming girl 游泳的女孩（分词）

a swimming pool 游泳的池子（动名词）

2) 分词作状语：

e.g. Seen from the top of the mountain, the lake looks like a tortoise. 从山顶上看，这个湖像一只乌龟。（过去分词和句子主语之间表示被动的关系）

e.g. Walking through the park, I found many beautiful flowers. 走过公园，我看见许多漂亮的花。（现在分词和句子主语之间表示主动的关系）

3) 分词作表语：

e.g. The story is interesting. 这个故事很有趣。（主语是物体，用现在分词作表语）

I am interested in computer games. 我对电脑游戏感兴趣。（主语是人，用过去分词作表语）

4) 分词作宾语补足语：可以跟宾语补足语的谓语动词有 see, watch, hear, set, keep, find, have, get 等。

e.g. I saw him walking in the street. 我看见他在街上走。

Exercises

Choose the appropriate answer from three choices marked A, B and C.

1. He told me something ___________ last night.
   A. to happen　　B. happened　　C. happening
2. The film was so ___________ that all of us were ___________ to tears.
   A. moving; moving　　B. moving; moved　　C. moved; moving
3. I found my son quite ___________ in swimming.
   A. interest　　B. interesting　　C. interested
4. With the job ___________, they went to traveling.
   A. doing　　B. done　　C. to do
5. I had my homework ___________ yesterday.
   A. to finish　　B. finishing　　C. finished
6. When I entered the classroom, I saw him ___________ in the first row.
   A. sitting　　B. sat　　C. seat

Part Two　Writing

Application form 申请表

表格的形式很多，常见的有留言表、个人信息表、申请表等，今天来看看加班申请表的填写。加班申请表包括的内容有：申请人、部门、时间、加班时间、加班原因等。

Sample

Overtime Request Form

Request Date: Nov. 10, 2019 ________

Employee's Name: Zhang Zhiqiang ____________

Department: Customer Service Department ____________

Date of Overtime: Nov.11, 2019 ____________

Overtime needed: from 6 o'clock p.m. to 12 o'clock p.m.

Total Overtime: not to exceed 6 hours

Reasons for Overtime Required: Double 11th is coming. People's online shopping will sharply increase on that day and the need for after-sales service will increase too.

____

☞ **Task**

说明：假定你是生产部的员工张建新，请根据下列内容填写一份加班申请表。

申请日期：2018 年 6 月 15 日

部门：生产部（Product Department）

加班时间：2018 年 6 月 20 日 9:00 a.m.— 4:00 p.m.

总加班时间：不超过 7 小时

加班原因：最近用电量增加，为保证居民和商业用电的稳定，公司需要加大发电量，增加了一台机组的运行，需安排人员对设备进行监管和维护。

Overtime Request Form

Request Date: (1) ____________

Employee's Name: (2) ____________

Department: (3) ____________

Date of Overtime: (4) ____________

Overtime needed: from (5) ____________ to (6) ____________

Total Overtime: not to exceed (7) ____________ hours

Reasons for Overtime Required: (8) ____

____

____

____

Section D An English Song

☞ **Task** Listen to the song and fill in the blanks with the missing words you have just heard, and then sing along with it.

I Need Some Sleep

Eels

I need some sleep 我要睡会儿了
You can't go on like this 不能再继续这样了
① ______________________________

But there's one I always miss 却总是事与愿违
Everyone says, "I'm getting down too low" 每个人都对我说"你陷得太深了"
Everyone says, "You just gotta let it go" 每个人都对我说"还是让这段感情过去吧"
You just gotta let it go 还是让这段感情过去吧
You just gotta let it go 还是让这段感情过去吧
② ______________________________

Time to put the old horse down 让疲惫的心得以休息
③ ______________________________

And the wheels keep spinning round 爱的惯性让我无法停止
Everyone says, "I'm getting down too low" 每个人都对我说"你陷得太深了"
Everyone says, "You just gotta let it go" 每个人都对我说"还是让这段感情过去吧"
You just gotta let it go 还是让这段感情过去吧
④ ______________________________

Background Tip:

Eels 是一支美国独立乐队，由三个成员组成，20 世纪 90 年代初出道。Eels 跟别的乐队相比有着非常不同的气质，拒绝过分商业化和低调是他们的特色，其作品偏向冷色调。*I Need Some Sleep* 这支单曲收录在专辑 *Shrek 2* 中。

Unit 7

Transmission and Distribution Systems

输配电系统

Section A Professional Background

Part One Lead-in

☞ **Task** Study the pictures and discuss the questions below.

Questions:

1. What can you see in the pictures?
2. What are the devices used to do?

Cues:
tower, transmission line, overhead line, substation, device, transformer, line, insulator, tension insulator string, concrete pole

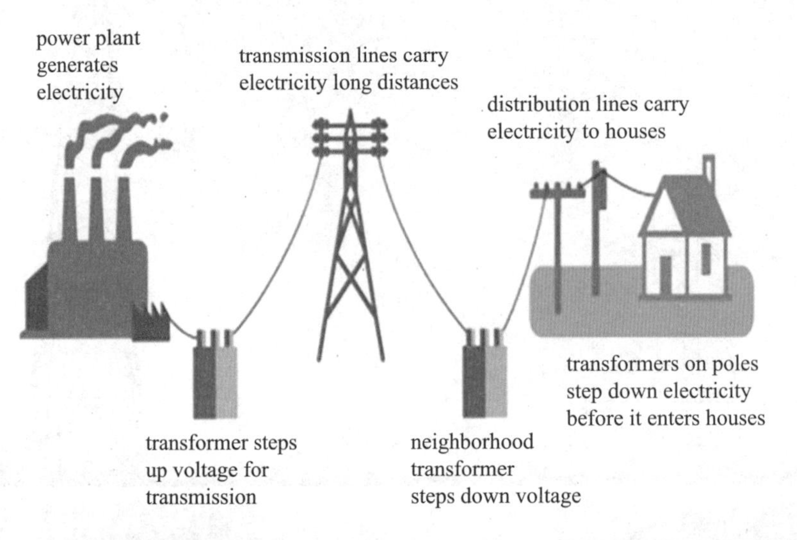

Part Two　Listening & Speaking

Listen to the following dialogues and fill in the blanks.

Dialogue 1

W: Hey, Tom, what are you __________ these days?

M: The exams are __________.

W: What exams will you take?

M: I will take the exams of High-Voltage Certificate and Aerial Work Certificate.

W: Are these exams __________?

M: I think they are, so I have to __________ for it.

W: Oh, I hope you can __________ these exams.

M: Thank you.

Dialogue 2

W: Hey Matthew, do you know what is __________ in the southeast of the south garden living area?

M: The Training Base for Transmission and Distribution Lines (输配电线路实训基地).

W: Tell me more about it.

M: In the training base, you can __________ the concrete pole, the tower, the transmission and distribution lines and other training facilities.

W: How many training projects will you take?

M: We will take twelve training projects.

W: What are they?

M: Transmission Line Design, Power Distribution Lines Construction, Pole-mounted Transformer Installation (柱上变压器安装), Transmission Line Measurement and so on.

W: That sounds cool.

M: If you are __________ in it, I can show you around the base and introduce for you.

W: Oh, thank you. I am __________ to your introduction.

M: You are welcome.

Section B Text Learning

Part One Lead-in

1. How many kinds of electric tower (杆塔) do you know?

2. How can the electricity be transmitted and distributed to users?

Part Two

Text A

Layout of Transmission and Distribution Systems

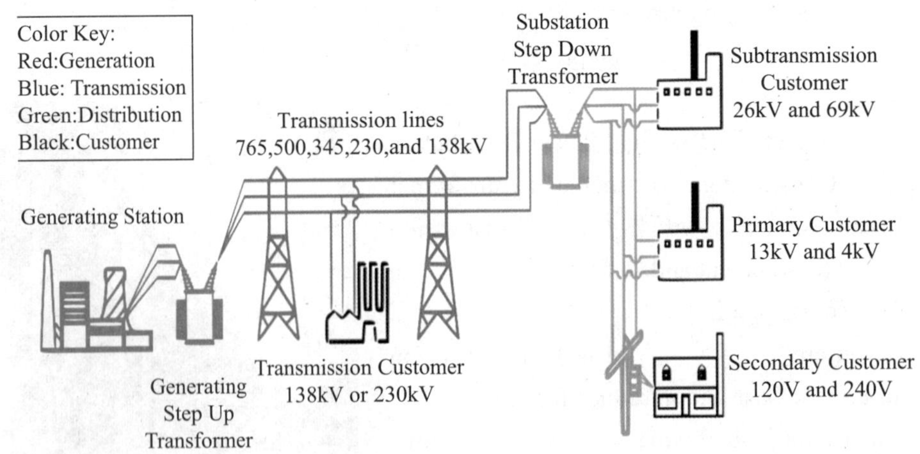

Fig. 7.1

Diagram of an electric power system: power generation is in red; transmission system is in blue; distribution is in green; customers are in black.

It is clear from Fig. 7.1 that bulk amount of power is generated at the power plant. The voltage which comes from the power plant is 11 kV. Then power goes to the nearby substation where the voltage is stepped up from 11 kV to 800 kV~220 kV. The power is then carried at such high voltage over long distance with the help of wires called transmission lines. Most transmission lines are high-voltage three-phase alternating current (AC), although single phase AC is sometimes used in railway electrification systems. High-voltage direct-current (HVDC) technology is used in long-distance transmission for its greater efficiency (typically hundreds of miles).

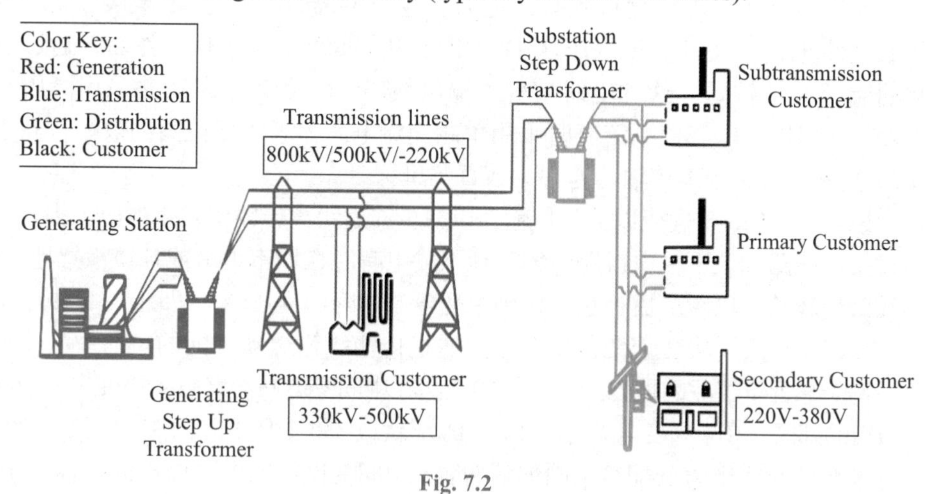

Fig. 7.2

Electricity is transmitted at high voltages (220 kV or above) to reduce the power loss which occurs in long-distance transmission. Electricity can be transmitted through overhead lines and underground transmission. Underground transmission is sometimes used in urban areas or environmentally sensitive locations. When the transmission line has covered a major distance, it is connected to step-down transformer at the main substation (see Fig. 7.2), where the voltage is stepped down to 500 kV–330 kV. From this substation, power is carried through subtransmission lines to different load centers (see Fig. 7.2). The main substation is located at places near the heavy industries. The subtransmission lines terminate at another substation called intermediate substation that is located in the outskirts of a city or town. The power is stepped down from 500 kV–330 kV to 330 kV/220 kV at intermediate substation.

The power is distributed to various consumers according to distribution lines. The power from the intermediate substation can be supplied to light industries. The domestic and industrial voltages are at 220 V and 380 V respectively, therefore, the power from the intermediate substation at 330 kV/220 kV is supplied to 220 kV–110 kV transformers installed at city substation.

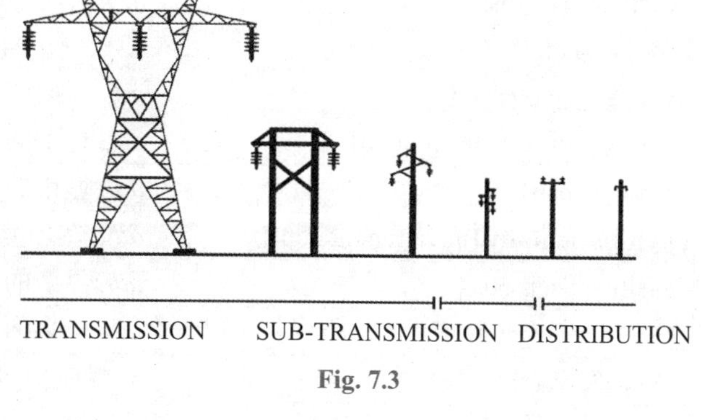

Fig. 7.3

(319 words)

参考译文

输配电系统布局

电力系统图：红色部分是发电；蓝色部分是电力传输系统；绿色部分是配电系统；黑色部分是电力用户。

从图中可以看出，电厂生产大量的电力。电厂出来的电压是 11 千伏。然后电力输送到附近的变电站，通过变压电压从 11 千伏上升到 800–220 千伏。然后，通过电线把高压电进行远距离输送，这种电线被称为输电线路。大多数输电线路采用高压三相交流电（AC），但铁路电气化系统有时采用单相交流电。为了实现更高的效率，高压直流（HVDC）技术用于长距离传输（通常为数百英里）。

电力以高压（220 千伏或以上）传输，以减少在远距离传输中发生的电力损耗。电力可以通过架空线路和地下输电线路传输。地下输电有时在城市地区或环境敏感地区使用。当传输线跨越较长一段距离后，它连接到主变电站的降压变压器（见图），其中电压降至 500 千伏或 330 千伏。从这个变电站，电力通过传输线输送到不同的负荷中心（见图），主变电站位于靠近重工业的地方。二次传输线路在中间变电站的终点，该变电站位于一个城市或城镇的郊区。在中间变电站，电压从 500 千伏或 330 千伏下降到 330 千伏 – 220 千伏。

根据配电线路将电力分配给不同的消费者。中间变电站的电可以提供给轻工业用电。家庭和工业的用电电压分别为 220 伏和 380 伏，因此，从中间变电站出来的 220 千伏电传输到城市变电站，此变电站安装有 110–220 千伏变压器。

Words & Expressions

diagram ['daɪəgræm]	*n.* 图解；图表；示意图
customer ['kʌstəmə]	*n.* 顾客
bulk [bʌlk]	*n.* 体积；容积；大块；大部分
wire ['waɪə(r)]	*n.* 金属丝；电线
electrification [ɪ'lektrɪfɪ'keɪʃn]	*n.* 电气化；带电
efficiency [ɪ'fɪʃnsi]	*n.* 效率；功率
reduce [rɪ'dju:s]	*v.* 减少；缩小
overhead [ˌəʊvə'hed]	*adj.* 高架的 *adv.* 在空中；在高处
underground [ˌʌndə'graʊnd]	*adj.* 地下的 *n.* 地下；地铁
urban ['ɜ:bən]	*adj.* 城市的；都市的
environmentally [ɪnˌvaɪrən'mentəli]	*adv.* 在环境方面地
sensitive ['sensətɪv]	*adj.* 灵敏的；敏感的
major ['meɪdʒə(r)]	*adj.* 较多的；主要的 *n.* 主修（科目）*v.* 主修
cover ['kʌvə(r)]	*n.* 封面 *v.* 覆盖；涉及；包含
connect [kə'nekt]	*v.* 连接；接通
substation ['sʌbsteɪʃn]	*n.* 分局；变电所

load center	负载中心，载荷中心
terminate ['tɜːmɪneɪt]	*v.* 结束；终止；满期；达到终点
domestic [də'mestɪk]	*adj.* 家庭的；国内的
install [ɪn'stɔːl]	*v.* 安装，装置，设置

Notes

1. 在中国，家庭用电电压为单相 220V，工业用电一般为三相 380V。
2. alternating current (AC): 交流电　direct current (DC): 直流电
3. underground transmission: 地下输电
4. be located at: 位于……位置
   e.g. Our campus is located at the suburb of the city.
5. be distributed to: 分给……；被分配到；分布

Exercises

☞ **Task 1**　Answer the following questions according to the text.

1. What is the voltage coming from the power plant?
   ____
2. Why is electricity transmitted at high voltages (138 kV or above) over long-distance transmission?
   ____
3. In what way can electricity be transmitted?
   ____
4. Where is the main substation located at?
   ____
5. What are domestic and industrial voltages?
   ____

☞ **Task 2**　Change the word formation.

Before	After	Meaning
efficiency　*n.*		*adj.* 效率高的；胜任的
reduce　*v.*		*adj.* 减少的；简化的
environmentally　*adv.*		*adj.* 环境的；有关环境的
sensitive　*adj.*		*adv.* 易感知地；神经过敏地
install　*v.*		*n.* 安装；装置

☞ **Task 3** Complete the following sentences with the proper form of the words from Task 2.

1. The worker is ___________ the pole.
2. This ___________ effect of this new thermal power plant could be serious.
3. The factory has built a highly ___________ dust removal system(除尘系统).
4. During working, his hearing is very ___________ .
5. The electricity supply has been ___________ this month.

☞ **Task 4** Translate the following English sentences into Chinese.

1. The power is then carried at such high voltage over long distance with the help of wires called transmission lines.

____

2. Electricity can be transmitted through overhead lines and underground transmission.

____

3. When the transmission line has covered a major distance, it is connected to step-down transformer at the main substation (See Fig. 7.2), where the voltage is stepped down to 500 kV–330 kV.

____

4. The main substation is located at places where the heavy industries are nearby.

____

5. The domestic and industrial voltages are at 240 V and 380 V respectively, therefore, the power from the intermediate substation at 330 kV/220 kV is supplied to 220 kV–110 kV transformers installed at city substation.

____

Part Three

Text B

Ultra-High Voltage Transmission Breaks China's Cycle of Energy Dependence

How to meet the energy need while also reducing pollution is one of the most urgent dilemmas facing the world. A technology used in China aims to do exactly that.

The technology is called ultra-high voltage transmission (or UHV), and it tackles problems that have long perplexed the energy industry. First, how to absorb the highly fluctuated power generation from renewable energy sources such as wind and solar. Second, how to transmit the electricity from remote areas to consumption centers without big losses over distances. At present, energy sources such as coal fired power stations need to be placed near cities to power them—helping pollute the air as they drive the economy.

Solving these challenges is critical to saving lives, lifting millions of people out of polluted living environment and minimizing climate change. Today we are over reliant on energy linked to

significant air pollution and to respiratory illnesses such as asthma. Cities depend on established energy sources that are closer to home.

UHV offers China a chance to break this cycle of dependence. It overcomes the problem of energy losses by increasing the voltage of transmission lines to 800 kV or even 1,000 kV.

So far we have been able to run effective transmission over lines as long as 2,500 kilometers and further lines at up to 5,000 kilometers are planned. This means we can see a future with less pollutants in high population areas. What's more, it means we can seriously consider bringing solar, wind and hydro into the main grid at scale.

(254 words)

参考译文

超高压输电可以打破中国能源依赖的循环

如何在满足能源需求的同时减少污染是目前全世界面临的共同难题之一。中国当前使用的一项技术正是以此为目标。

这项技术被称为超高压输电（简称 UHV），它解决了长期以来困扰能源产业的难题。第一，如何吸收风能、太阳能等可再生能源产生的剧烈波动的电力。第二，如何将电力从偏远地区输送到消费中心，而不会造成由于远距离传输而产生巨大的损耗。目前，为了供电，像燃煤发电站这样的能源需要放置在城市附近——它们推动了经济发展的同时也污染了空气。

解决这些挑战对于拯救生命、使数百万人摆脱受污染的生活环境和减少气候变化至关重要。今天，我们过度依赖于严重污染空气以及会引起哮喘等呼吸道疾病有关的能源。城市依赖于离家较近的现有能源。

超高压为中国提供了打破这种依赖循环的机会。它通过将输电线路电压提高到 800 kV 甚至 1 000 kV 来解决能量损失问题。

到目前为止，我们已经能够用线路进行长达 2 500 公里的有效传输，并计划达到 5 000 公里以内的线路传输。这意味着，我们将来可以看到，在人口高度集中的地区，污染物变少了。更令人兴奋的是，这意味着我们真的可以考虑将太阳能、风能和水力发电纳入电网的规模。

Words & Expressions

dilemmas [dɪ'leməz]	*n.* 困境，窘境（名词 dilemma 的复数形式）
perplex [pə'pleks]	*v.* 使困惑；使糊涂；使复杂化
fluctuate ['flʌktʃueɪt]	*vi.* 波动；动摇；起伏 *vt.* 使动摇
remote [rɪ'məʊt]	*adj.* 遥远的；偏僻的
minimize ['mɪnɪmaɪz]	*v.* 将…… 减到最少
reliant [rɪ'laɪənt]	*adj.* 依赖的
respiratory [rə'spɪrətri]	*adj.* 呼吸的；与呼吸有关的
asthma ['æsmə]	*n.* 哮喘

Notes

ultra-high voltage transmission: 超高压输电
respiratory illnesses: 呼吸道疾病

Exercises

☞ **Task 1** The following picture is a ultra-high voltage transmission tower. Fill in the blanks with the right words or phrases according to the marks in the picture.

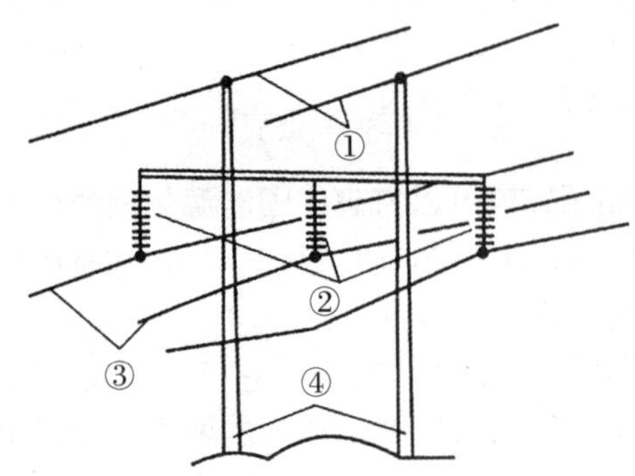

① ______________________________
② ______________________________
③ ______________________________
④ ______________________________

☞ **Task 2** Match the words with their meanings.

optimal	排放
emission	将……减到最少
minimize	依赖的
reliant	最理想的
link	联系；连接

☞ **Task 3** Translate the following sentences into Chinese.

1. The technology is called ultra-high voltage-transmission (or UHV), and it tackles problems that have long perplexed the energy industry.

____

____

2. Solving these challenges is critical to saving lives, lifting millions of people out of polluted living environment and minimizing climate change.

____

____

3. Cities depend on established energy sources that are closer to home.

____

____

4. It overcomes the problem of energy losses by increasing the voltage of transmission lines to 800 kV or even 1,000 kV.

____

____

5. We can see a future with less pollutants in high population areas.

____

____

Section C　Advanced Training

Part One　Grammar: 动词不定式

动词不定式是动词的一种非限定形式，而非限定性动词是指动词不受主语限制的形式。非限制性动词指动词不定式、动词 -ing 和动词 -ed 的形式。它们不可以独立地充当句子的谓语动词，所以也不受主语的人称和数的制约。

1. 不定式的时态和语态

时态 \ 语态	主动	被动
一般时	to do	to be done
进行时	to be doing	
完成时	to have done	to have been done
完成进行时	to have been doing	

1) 现在时：一般现在时表示的动词，有时与谓语动词表示的动作同时发生，有时发生在谓语动词的动作之后。

e.g. I hope to replace the wiring as it is faulty. 因为线路故障，我想更换线路。

2) 完成时：表示的动作发生在谓语动词表示的动作之前。

e.g. It is thrilling that the company has built an ultra-high-voltage cross-country transmission network.（动词的完成时态）

那个公司已经建成了跨越全国的超高压输电网络，这太令人兴奋了。

I'm sorry to have given you so much trouble.（不定式的完成时态）

很抱歉给你带来那么多麻烦。

3) 进行时：表示动作正在进行，与谓语动词表示的动作同时发生。

I see she is working on the new project.（动词的现在进行时）

我看见她正忙于那个新项目。

The facilities seem to be operating normally.（不定式的进行时）

那些设备似乎运转正常。

4) 完成进行时：表示在谓语动作之前一直在进行的动作。

He is known to have been studying ultra-high-voltage technology for 5 years.

他致力于研究超高压技术已经 5 年了，并因此而闻名。

2. 动词不定式在句中可以作主语、宾语、表语、定语或状语。

1) To get to the power station by subway is about half an hour.（作主语）

坐地铁到电厂大约需要半个小时。

e.g.It is easy to get to the power station by subway.

坐地铁去电厂很便利。

2) I want to practice in the wind power station.（不定式作宾语）

我想要去风力发电站实习。

[类似的动词有 afford（付得起），agree（同意），aim（力求做到），ask（要求），attempt（试图），choose（决定），decide（决定），demand（要求），expect（期待），hope（希望），learn（学会），manage（设法），fail（失败），offer（主动提出），plan（计划），prepare（准备），pretend（假装），seem（觉得好像），tend（往往会），want（想要），wish（希望）等]

e.g.I don't know how to overcome the technical puzzles. [疑问词＋不定式 how/ what/ where/ when] 我不知道如何才能攻克那些技术难关。

3) His job is to examine the transformers.（作表语）

他的工作是检验变压器。

4) I have nothing to tell you about the accident.（作定语）

关于这次事故，我无可奉告。

5) I learn the major of power station and power system in order to work in the electric power system.（作状语，可表示目的、原因、结果、方式等）

我学发电厂及电力系统专业是为了以后进电力系统工作。

Exercises

Choose the appropriate answer from three choices marked A, B and C.

1. The moderator asked us ___________ so much noise.
   A. don't make　　B. not make　　C. not to make

2. ___________ is a requirement for employees in our company.
   A. To practice　　B. Practice　　C. Being practiced

3. The meeting ___________ is at 10 a.m.
   A. to take place　　B. to be taken place　　C. taking place

4. I failed ___________ my presentation during the seminar.
   A. and finished　　B. to finish　　C. to finished

5. I feel it an honor ___________ to speak here.
   A. to be asked　　B. to ask　　C. asked

6. Will you lend me the book ___________?
   A. to be read　　B. to read　　C. be read

Part Two　Writing

Resume 简历

简历即个人履历，是求职时用来向聘方介绍自己的书面报告。简历要写得具体、简洁

明了、一目了然。

写简历应该包含以下几点内容：

工作经历（Work experience）。列出曾从事的工作，要按顺序陈列清楚从事每项工作的时间、职衔及职责。

教育状况（Education）。突出与求职的工作紧密相关的课程。

个人信息（Personal information）。个人信息包括以下内容：

姓名（Name）：Family name (last name) 是姓，Given name (first name) 是名，西方人是名在前，姓在后；而中国人是姓在前，名在后。

性别（Gender/Sex）：男性用 Male，女性用 Female。

出生年月（Date of birth）：如 “Dec12, 1998”。

地址（Address）：英文的顺序是从小到大如 No.39 Keyuan Road, Xixiangtang District, Nanning, Guangxi。

婚姻状况（Marital status）：已婚用 married，未婚用 single。

国籍（Nationality）：中国国籍用 Chinese，而不用 China。

联系电话（Telephone）。

电子邮件地址（E-mail address）。

根据下列内容写一篇英文简历。

姓名：李艾华，男，1995 年 5 月 16 日出生，未婚，家住北京市复兴路 61 号。2015 年毕业于广西电力职业技术学院，成绩优秀。2015 年 6 月至 2017 年 7 月在麻石电厂工作，2017 年 7 月至 2018 年 12 月在来宾电厂工作，工作岗位是电工。

Sample

<table>
<tr><td colspan="2">Resume
Personal information</td></tr>
<tr><td>Name</td><td>Li Aihua</td></tr>
<tr><td colspan="2">Date of birth: May 16, 1995</td></tr>
<tr><td>Gender/Sex</td><td>Male</td></tr>
<tr><td colspan="2">Marital status: single</td></tr>
<tr><td colspan="2">Address: 61 Fuxing Road, Beijing</td></tr>
<tr><td colspan="2">Zip Code: 100036</td></tr>
<tr><td colspan="2">E-mail address: 752634@sina.com</td></tr>
<tr><td colspan="2">Telephone:17233336665</td></tr>
<tr><td colspan="2">Education: Graduated from the Guangxi Electrical Polytechnic Institute in 2015 with excellent scores.</td></tr>
<tr><td colspan="2">Work expression:
2017.07-2018.12: working in Laibin Power Plant as an electrician.
2015.06-2017.07: working in Mashi Power Plant as an electrician.</td></tr>
</table>

☞ **Task**

根据下列内容写一篇英文简历。

姓名：雷原，男，1998 年 8 月 15 日出生，未婚，家住南宁市中山路 20 号。2018 年 7 月毕业于广西电力职业技术学院，高中就读于南宁市 13 中。大学毕业至今一直在广西西津水力发电厂工作，职位是发电技术员。

Resume	
Personal information	
Name	
Date of birth:	
Gender/Sex	
Marital status:	
Address:	
Zip Code:	
E-mail address:	
Telephone:	
Education:	
Work expression:	

Section D An English Song

☞ **Task** Listen to the song and fill in the blanks with the missing words you have just heard, and then sing along with it.

Someone's Watching Over Me

Hilary Erhard Duff

Found myself today 我找到自我，今天
Oh I found myself and ran away 找到了，也逃避了
Something pulled me back 被莫名地拉回来
The voice of reason I forgot I had 忘记了曾经拥有的理性
All I know is just you're not here to say 只知道你不在了
What you always used to say 听不到你常说的那些话了
But it's written in the sky tonight 但是今晚这些话赫然写在夜空之上
① ________________

No I won't break down 不会消沉

Sooner than it seems life turns around 我知道转眼间，生命自有出路
And I will be strong 而我也会变坚强
Even if it all goes wrong 尽管也许世事都不如意
When I'm standing in the dark I'll still believe 人生低潮时我会始终相信
Someone's watching over me 有人在守护着我
Seen that ray of light 看见了那道光
And it's shining on my destiny 它照亮我的人生
Shining all the time 一直都在
And I won't be afraid 我因此不再害怕
To follow everywhere it's taking me 跟随它的指引
All I know is yesterday is gone 我只知道过去已成往事
② ______________________________

Took this moment to my dreams 珍惜当下，实现梦想
So I won't give up 所以我不会放弃
No I won't break down 不会消沉
Sooner than it seems life turns around 我知道转眼间，生命自有出路
③ ______________________________

Even if it all goes wrong 尽管世事都不如意
When I'm standing in the dark I'll still believe 人生低潮时我始终相信
Someone's watching over me 有人在守护着我
It doesn't matter what people say 别人怎说都不重要
And it doesn't matter how long it takes 花多长时间也没关系
Believe in yourself and you'll fly high 相信自己，就可以高飞
④ ______________________________

Be true to yourself and follow your heart 真诚面对自己，跟随自己的心意

Background Tip:

Hilary Erhard Duff，出生于 1987 年 9 月 28 日，是美国演员、歌手、企业家、设计师，被誉为“美国甜心”。*Someone's Watching Over Me* 是一首非常励志的歌曲。这首歌出自 Hilary 参与主演的电影《劲歌飞扬》（*Raise Your Voice*）。

Glossary

A		
add [æd]	v. 添加；增加；补充说；继续说	U6A
agricultural [ˌægrɪ'kʌltʃərəl]	adj. 农业的；农艺的	U5B
air-preheater ['eəpri:hi:tə]	n. 空气预热器	U4A
apparatus [æpə'reɪtəs]	n. 设备；装置；仪器	U1A
appliance [ə'plaɪəns]	n. 器具；器械；装置；应用	U3B
appropriate [ə'prəʊprɪət]	adj. 适当的；相称的	U4A
asthma ['æsmə]	n. 哮喘	U7B
atmospheric pollution	大气污染	U4B
attach [ə'tætʃ]	v. 附上；系上；贴上	U4A
automate ['ɔ:təmeɪt]	v. 使自动化	U2B
availability [əˌveɪlə'bɪləti]	n. 有效；有用；可用性	U2B
B		
bash [bæʃ]	v. 猛击，痛击；严厉批评	U3B
batch [bætʃ]	n. 一批；一炉；一次所制之量	U6B
battle ['bætl]	n. 战争，战役	U1B
beam [bi:m]	n. 梁；光线；平衡木；（电波的）波束 v. 照射；发光	U6A
Belt and Road	“一带一路”	U1B
boiler ['bɔɪlə(r)]	n. 锅炉；汽锅；热水器	U3A
boost [bu:st]	vt. 促进；增加；支援 vi. 宣扬	U4B
bulk [bʌlk]	n. 体积；容积；大块；大部分	U7A
burn [bɜ:n]	vt. 燃烧；烧毁，灼伤；激起……的愤怒 vi. 燃烧；烧毁；发热	U4B
C		
capacity [kə'pæsəti]	n. 容量；容积；能力；职位	U5A
crash [kræʃ]	v. 碰撞；（使）摔碎；（机器、系统等）崩溃	U3B
cell [sel]	n. 元件；【生】细胞；【电】电池	U6A
central receiver	中央接收器	U6A
chief researcher	首席研究员	U6B
circuits ['sɜ:kɪts]	n. 电路（circuit 的复数形式）	U3A
circulation [ˌsɜ:kjə'leɪʃn]	n. 流通；循环	U4A
combustion [kəm'bʌstʃən]	n. 燃烧	U3A U4A
community [kə'mju:nəti]	n. 社区；[生态]群落；共同体；团体	U5B

component [kəm'pəʊnənt]	*n.* 零部件；元件	U3A
comprised [kəmp'raɪzd]	*adj.* 包含的（动词 comprise 的过去式和过去分词）	U5A
computerized [kəm'pju:təˌraɪzd]	*adj.* 用计算机操作（管理）的	U2B
concentrate ['kɒnsəntreɪt]	*vi.* 集中；浓缩；全神贯注；聚集 *vt.* 集中；浓缩	U6B
condense [kən'dens]	*v.* 浓缩；凝结；缩短	U3A
condenser [kən'densə]	*n.* 冷凝器；电容器	U3A
connect [kə'nekt]	*v.* 连接；接通	U7A
consumption [kən'sʌmpʃ(ə)n]	*n.* 消费；消耗量	U6A
consumption [kən'sʌmpʃn]	*n.* 消费；消耗	U1B
conversion [kən'vɜ:ʃn]	*n.* 转变；换算	U1A
corridor ['kɒrɪdɔ:(r)]	*n.* 走廊	U3B
coupled ['kʌpld]	*adj.* 成对的（连接的；共轭的；联系的）	U3A U5A
cover ['kʌvə(r)]	*n.* 封面 *v.* 覆盖；涉及；包含	U7A
crash [kræʃ]	*v.* 碰撞；（使）摔碎；（机器、系统等）崩溃	U3B
current ['kʌrənt]	*n.*（水、气、电）流；趋势；涌流 *adj.* 现在的；流通的，最近的	U2B U3B
customer ['kʌstəmə]	*n.* 顾客	U7A
D		
dam [dæm]	*n.* 水坝；堤 *v.* 筑坝	U5A
demonstration [ˌdemən'streɪʃən]	*n.* 示范；证明；示威游行	U6B
depend on	取决于；依赖；依靠	U5B
design [dɪ'zaɪn]	*n.* 设计；图样 *v.* 设计	U4A
destination [ˌdestɪ'neɪʃn]	*n.* 目的地，终点	U3B
diagram ['daɪəgræm]	*n.* 图解；图表；示意图	U7A
digital ['dɪdʒɪtl]	*adj.* 数字的；数码的	U2B
dilemmas [dɪ'leməz]	*n.* 困境，窘境（名词 dilemma 的复数形式）	U7B
dish-engine	碟式太阳能发电系统	U6A
dismantle [dɪs'mæntl]	*vt.* 拆除；取消；解散	U5B
dissolve [dɪ'zɒlv]	*vt.* 使溶解；使分解 *vi.* 溶解；解散	U5B
distribution system	配电系统	U2A
disturbance [dɪ'stɜ:bəns]	*n.* 扰乱；骚动	U2B
domestic [də'mestɪk]	*adj.* 家庭的；国内的	U7A
dry out	变干；戒酒	U5B
E		
ecological [ˌi:kə'lɒdʒɪkl]	*adj.* 生态（学）的	U1B
economizer [ɪ'kɒnəˌmaɪzə]	省煤器	U4A

efficiency [ɪ'fɪʃnsi]	*n.* 效率；功率	U7A
electric motor	电动机	U1A
electricity [ɪˌlek'trɪsəti]	*n.* 电	U1A
electrification [ɪ'lektrɪfɪ'keɪʃn]	*n.* 电气化；带电	U7A
electromagnetic [ɪˌlektrəʊmæg'netɪk]	*adj.* 电磁的	U3A
electron [ɪ'lektrɒn]	*n.* 电子	U3B
electronics [ɪˌlek'trɒnɪks]	*n.* 电子学；电子器件	U1A
emission [ɪ'mɪʃən]	*n.* 排放；排放物；（光、热等的）发射	U5B
emission [ɪ'mɪʃən]	*n.*（光、热等的）发射，散发；排放；发行	U4B
emit [ɪ'mɪt]	*vt.* 发出，放射；发行；发表	U4B
environmentally [ɪnˌvaɪrən'mentəli]	*adv.* 在环境方面地	U7A
equip [ɪ'kwɪp]	*vt.* 装备；具备 *n. abbr.* 装备（=equipment）	U4A
exhaust [ɪg'zɔːst]	*v.* 耗尽 *n.* 排气装置；废气	U3A
expansion [ɪk'spænʃn]	*n.* 扩大；扩张	U1B
extremely [ɪk'striːmli]	*adv.* 非常，极其；极端地	U3B
F		
facility [fə'sɪləti]	*n.* 设施；设备	U5B
feed [fiːd]	*vt.* 喂养；饲养；向……提供 *n.* 饲料；饲养	U2A U4A
flow [fləʊ]	*n.* 流动；流；流量 *vi.* 流动；涌出；飘动 *vt.* 淹没	O3B U5A
fluctuate ['flʌktʃʊeɪt]	*vi.* 波动；动摇；起伏 *vt.* 使动摇	U7B
focus ['fəʊkəs]	*n.* 关注；调焦；中心点（指人或事物） *v.* 集中（注意力、精力等于）；（使）调节焦距；集中（光束于）	U6A
fossil ['fɒs(ə)l]	*n.* 化石；老人；老古董 *adj.* 从地下发掘出来的；化石的，fossils（复数）	U6A
fossil fuels	化石燃料	U6B
furnace ['fɜːnɪs]	*n.* 炉子；炉膛	U4A
G		
generating center	发电中心	U2A
generating station	发电站	U2A
generator ['dʒenəreɪtə]	*n.* 发电机；发生器	U1A
geothermal power	地热能	U1A
governance ['gʌvənəns]	*n.* 统治；管理；支配；统治方式	U1B
gravity ['grævəti]	*n.* 重力；庄重	U5A
grid [grɪd]	*n.* 电网；网格	U2A
groundswell ['graʊndswel]	*n.* 风潮；暴涌	U2B

H		
habitat ['hæbɪtæt]	*n.* [生态] 栖息地，产地	U5B
handling ['hændlɪŋ]	*n.* 处理；操作	U4A
heavy metal	重金属	U4B
height [haɪt]	*n.* 高度；高处；顶点	U4A U5A
heliostat ['hiːlɪə(ʊ)stæt]	*n.* [天] 定日镜；日光反射装置	U6A
hinder ['hɪndə]	*vi.* 成为阻碍 *vt.* 阻碍；打扰	U6B
hydraulic [haɪ'drɔːlɪk]	*adj.* 水力的；水压的；液压的	U5A
hydropower plant	水电站	U1A
I		
induction [ɪn'dʌkʃn]	*n.* 感应；诱发	U3A
inexhaustible [ˌɪnɪg'zɔːstəb(ə)l]	*adj.* 用之不竭的；无穷无尽的	U6A
install [ɪn'stɔːl]	*v.* 安装，装置，设置	U7A
installed	install 的过去式和过去分词	U6A
intake ['ɪnteɪk]	*n.* 入口；通风口；吸入；招收	U5A
integral ['ɪntɪgrəl]	*adj.* [数学] 积分的；完整的，整体的	U6B
integration [ˌɪntɪ'greɪʃn]	*n.* 集成；综合；同化	U2B
K		
kinetic [kɪ'netɪk]	*adj.* 运动的；动力学的	U3A U4A
kinetic energy	动能	U4A
L		
lens [lenz]	*n.* 透镜；镜片	U6A
load center	负载中心，负荷中心	U2A U7A
located [ləʊ'keɪtɪd]	*adj.* 处于；位于	U5A
M		
major ['meɪdʒə(r)]	*adj.* 较多的；主要的 *n.* 主修（科目）*v.* 主修	U7A
make sense	有意义；讲得通；言之有理	U3B
make way for	让路；为……开路	U5B
mechanical [mə'kænɪkl]	*adj.* 机械的；力学的	U3A
mechanical energy	机械能	U4A
metal ['metl]	*n.* 金属；合金 *adj.* 金属制的	U3B
minimize ['mɪnɪmaɪz]	*v.* 将……减到最少	U7B
minimize ['mɪnɪmaɪz]	*vt.* 使减到最少；小看 *vi.* 最小化	U5B
mirror ['mɪrə(r)]	*n.* 镜；反射镜；反映；借鉴 *v.* 反映；反射	U6A

N		
negative ['negətɪv]	*adj.* 负的；消极的；否定的；阴性的	U5B
non-fossil fuel energy	非化石能源	U6A
normally ['nɔːməli]	*adv.* 通常；正常地	U3A
nuclear power plant	核电厂	U1A
O		
occupy ['ɒkjʊpai]	*vt.* 占据，占领；居住；使忙碌	U6B
organism ['ɔːgənɪzəm]	*n.* 有机体；生物体；微生物	U5B
overhead [ˌəʊvə'hed]	*adj.* 高架的 *adv.* 在空中；在高处	U7A
oxygen ['ɒksɪdʒən]	*n.* [化学] 氧气，氧	U5B
P		
participation [pɑːˌtɪsɪ'peɪʃn]	*n.* 参加，参与；分享	U1B
particulate matter	微颗粒物质	U4B
peak [piːk]	*n.* 尖端；顶峰	U2B
penstock ['penstɒk]	*n.* 水门；水道；水闸	U5A
perplex [pə'pleks]	*v.* 使困惑；使糊涂；使复杂化	U7B
photocopier ['fəʊtəʊkopɪə(r)]	*n.* 复印机；影印机	U3B
photovoltaic [ˌfəʊtəʊvɒl'teɪk]	*adj.*【物】光电的 *n.* 光伏；太阳光电	U6A
photovoltaics (PV)	太阳能光伏发电	U6A
pollutant [pə'luːtənt]	*n.* 污染物	U4B U7B
pollution prevention	污染防治	U1B
possess [pə'zes]	*vt.* 拥有；持有；支配	U5A
potential [pə'tenʃl]	*adj.* 潜在的 *n.* 潜力；电位；电势	U5A
powerhouse ['paʊəhaʊs]	*n.* 发电站；动力室	U5A
preferential policy	优惠政策	U6B
promote [prə'məʊt]	*vt.* 促进，推进；提升	U1B
pulverized ['pʌlvəraɪzd]	*adj.* 毁坏的（动词 pulverize 的过去式及过去分词形式）	U4A
R		
reduce [rɪ'djuːs]	*v.* 减少；缩小	U7A
reflect [rɪ'flekt]	*vt.* 反映；反射，照出；反省 *vi.* 反射，映现；深思	U6B
release [rɪ'liːs]	*n.* 释放；发行 *vt.* 释放	U4A
reliant [rɪ'laɪənt]	*adj.* 依赖的	U7B
remote [rɪ'məʊt]	*adj.* 遥远的；偏僻的	U7B
renewable [rɪ'njuːəbl]	*adj.* 可更新的；可再生的	U2B
required [rɪ'kwaɪəd]	*adj.* 需要的	U4A

reservoir ['rezəvwɑː(r)]	*n.* 水库；储藏；蓄水池	U5A
residual [rɪ'zɪdjʊəl]	*adj.* 剩余的；残余的 *n.* 剩余部分	U4A
respiratory [rə'spɪrətri]	*adj.* 呼吸的；与呼吸有关的	U7B
restoration [ˌrestə'reɪʃn]	*n.* 恢复；归还；复位	U2B
revolution [ˌrevə'luːʃn]	*n.* 革命；彻底改变	U1B
rotate [rəʊ'teɪt]	*v.* 旋转；循环	U3A
rotor ['rəʊtə(r)]	*n.* 旋转体;〈机〉转子	U3A
	S	
sensitive ['sensətɪv]	*adj.* 灵敏的；敏感的	U7A
shaft [ʃɑːft]	*n.* 轴；柄；矛	U5A
shaver ['ʃeɪvə(r)]	*n.* 剃须刀	U3B
show [ʃəʊ]	*v.* 证明；显现；展示	U3A
significant [sɪg'nɪfɪkənt]	*adj.* 重大的；有意义的	U5B
smog [smɒg]	*n.* 雾霾	U4B
solar energy	太阳能	U6A
solar power	太阳能；太阳能动力	U1A
solar ['səʊlə(r)]	*adj.* 太阳的；太阳能的 *n.* 日光浴室	U6A
stack [stæk]	*n.*（一）堆，（一）叠；烟囱；许多，排气管 *v.* 堆积；堆放	U4A
steam turbine	汽轮机，蒸汽轮机	U2A
step-up	*adj.* 把电压升高的	U2A
step-up substation	升压变电站	U2A
storage ['stɔːrɪdʒ]	*n.* 贮藏；仓库	U4A
store [stɔː(r)]	*n.* 商店；贮藏 *vt.* 储存；贮藏；保存	U2A
substation ['sʌbsteɪʃn]	*n.* 分局；变电所	U7A
substructure ['sʌbstrʌktʃə]	子结构	U5A
superstructure ['suːpəstrʌktʃə(r)]	*n.*〈政〉上层建筑;〈建〉上层结构	U5A
	T	
tackle ['tækl]	*vt.* 处理；抓住；固定；与……交涉 *vi.* 擒抱摔倒；拦截抢球	U6B
tangible ['tændʒəbl]	*adj.* 有形的；切实的；可触摸的	U6B
terminate ['tɜːmɪneɪt]	*v.* 结束；终止；满期；达到终点	U7A
The Ministry of Science and Technology	科技部	U4B
The National Technology Invention Award	国家技术发明奖	U4B
The National Energy Administration	国家能源局	U6A

thermal power plant	热电厂	U1A
three-phase [θriː'feɪz]	*adj.* 三相的 *n.* 三相	U2A U7A
tidal power	潮汐能	U1A
tracking system	跟踪系统	U6A
transfer [træns'fɜː(r)]	*v.* 转移；调任；转乘 *n.* 迁移；移动；换车；汇兑	U2A
transformer [træns'fɔːmə(r)]	*n.* 变压器	U1A
transmission [træns'mɪʃn]	*n.* 传输；传播；变速器	U1A
turbine ['tɜːbaɪn]	*n.* 涡轮（机）	U3A
turn out	发生；结果是；出席；熄灭；生产，制造	U3B
	U	
ultralow [ˌʌltrə'ləʊ]	*adj.* 极低的	U4B
underground [ˌʌndə'graʊnd]	*adj.* 地下的 *n.* 地下；地铁	U7A
unprecedented [ʌn'presɪdentɪd]	*adj.* 空前的；前所未有的	U2B
urban ['ɜːbən]	*adj.* 城市的；都市的	U7A
utilize ['juːtəlaɪz]	*vt.*〈美〉利用或使用 =〈英〉utilise	U5A
	V	
vessel ['vesl]	*n.* 容器；器皿	U3A
vitalize ['vaɪtəlaɪz]	*vt.* 赋予生命；给予……生命；使有生气	U1B
voltage ['vəʊltɪdʒ]	*n.* 电压	U1A
	W	
wind power	风能	U1A
wire ['waɪə(r)]	*n.* 金属丝；电线	U7A

References

[1] https://www.mechanicalbooster.com/2016/08/steam-power-plant.html

[2] https://www.brighthubengineering.com/fluid-mechanics-hydraulics/7120-components-of-hydroelectric-power-plants-part-one/

[3] https://en.wikipedia.org/wiki/Concentrated_solar_power

[4] https://en.wikipedia.org/wiki/Electric_power_transmission

[5] https://www.smartgrid.gov/the_smart_grid/smart_grid.html)

[6] https://www.forbes.com/sites/realspin/2014/09/18/ultra-high-voltage-transmission-can-break-chinas-cycle-of-energy-dependence/#108cb98034f4

[7] http://www.chinadaily.com.cn/a/201907/12/WS5d27eaafa3105895c2e7d1d0.html, 2019-7-12/2019-7-24.

[8] https://www.chinadaily.com.cn/a/201907/23/WS5d3657dea310d830564006d0.html, 2019-7-23/2019-7-24.

[9] www.kekenet.com.

中国人民大学出版社外语出版分社读者信息反馈表

尊敬的读者：

感谢您购买和使用中国人民大学出版社外语出版分社的 ______________ 一书，我们希望通过这张小小的反馈卡来获得您更多的建议和意见，以改进我们的工作，加强我们双方的沟通和联系。我们期待着能为更多的读者提供更多的好书。

请您填妥下表后，寄回或传真回复我们，对您的支持我们不胜感激！

1. 您是从何种途径得知本书的：
   □书店 □网上 □报纸杂志 □朋友推荐
2. 您为什么决定购买本书：
   □工作需要 □学习参考 □对本书主题感兴趣 □随便翻翻
3. 您对本书内容的评价是：
   □很好 □好 □一般 □差 □很差
4. 您在阅读本书的过程中有没有发现明显的专业及编校错误，如果有，它们是：
   ____
   ____
   ____
5. 您对哪些专业的图书信息比较感兴趣：
   ____
   ____
   ____

6. 如果方便，请提供您的个人信息，以便于我们和您联系（您的个人资料我们将严格保密）：

您供职的单位：______________________________

您教授的课程（教师填写）：______________________________

您的通信地址：______________________________

您的电子邮箱：______________________________

请联系我们：黄婷　程子殊　吴振良　王琼　鞠方安

电话：010-62512737，62513265，62515538，62515573，62515576

传真：010-62514961

E-mail：huangt@crup.com.cn　chengzsh@crup.com.cn　wuzl@crup.com.cn
crup_wy@163.com　jufa@crup.com.cn

通信地址：北京市海淀区中关村大街甲 59 号文化大厦 15 层　邮编：100872

中国人民大学出版社外语出版分社